# SOLVENT EFFECTS IN CHEMISTRY

# SOLVENT EFFECTS IN CHEMISTRY

Second Edition

**ERWIN BUNCEL**
**ROBERT A. STAIRS**

Published by John Wiley & Sons, Inc., Hoboken, New Jersey
Published simultaneously in Canada

For general information on our other products and services or for technical support, please contact our Customer Care Department within the United States at (800) 762-2974, outside the United States at (317) 572-3993 or fax (317) 572-4002.

Wiley also publishes its books in a variety of electronic formats. Some content that appears in print may not be available in electronic formats. For more information about Wiley products, visit our web site at www.wiley.com.

*Library of Congress Cataloging-in-Publication Data*

Erwin Buncel
  Solvent effects in chemistry / Erwin Buncel, Robert A. Stairs.
    pages   cm
  Includes bibliographical references and index.
  ISBN 978-1-119-03098-0 (cloth)
1. Solvation.   2. Chemical reactions.   3. Solvents.   I. Buncel, E.   II. Title.
  QD543.S684 2015
  541′.34–dc23

                                          2015010522

Cover image courtesy of Professor Errol Lewars, Trent University.

Set in 10/12pt Times by SPi Global, Pondicherry, India

Printed in the United States of America

10  9  8  7  6  5  4  3  2  1

2  2016

# CONTENTS

# PREFACE TO THE SECOND EDITION

The present work is in effect the second edition of Buncel, Stairs, and Wilson's (2003) *The Role of the Solvent in Chemical Reactions*. In the years since the appearance of the first edition, the repertoire of solvents and their uses has changed considerably. Notable additions to the list of useful solvents include room-temperature ionic liquids, fluorous solvents, and solvents with properties "switchable" between different degrees of hydrophilicity or polarity. The use of substances at temperatures and pressures near or above their critical points as solvents of variable properties has increased. Theoretical advances toward understanding the role of the solvent in reactions continue. There is currently much activity in the field of kinetic solvent isotope effects. A search using this phrase in 2002 yielded 118 references to work on their use in elucidating a large variety of reaction mechanisms, nearly half in the preceding decade, ranging from the $S_N2$ process (Fang *et al.*, 1998) to electron transfer in DNA duplexes (Shafirovich *et al.*, 2001). Nineteen countries were represented: see, for example, Blagoeva *et al.* (2001), Koo *et al.* (2001), Oh *et al.* (2002), Wood *et al.* (2002). A similar search in 2013 yielded over 25,000 "hits."

The present edition follows the pattern of the first in that the introductory chapters review the basic thermodynamics and kinetics needed for describing and understanding solvent effects as phenomena. The next chapters have been revised mainly to improve the presentation. The most changed chapters are near the end, and attempt to describe recent advances.

Some of the chapters are followed by problems, some repeated or only slightly changed from the first edition, and a few new ones. Answers to most are provided.

We are grateful to two anonymous colleagues who reviewed the first edition when this one was first proposed, and who pointed out a number of errors and infelicities. One gently scolded us for using the term "transition state" when the physical entity, the activated complex, was meant. He or she is right, of course, but correcting it in a number of places required awkward circumlocutions, which we have shamelessly avoided (see also Atkins and de Paula, 2010, p. 844.). We hope that most of the remaining corrections have been made. We add further thanks to Christian Reichardt for steering us in new directions, and we also thank Nicholas Mosey for a contribution to the text and helpful discussions, and Chris Maxwell for Figure 5.11. We add David Poole, Keith Oldham, J. A. Arnot, and Jan Myland to the list of persons mentioned in the preface to the first edition who have helped in different ways. Finally, we thank the editorial staff at Wiley, in particular Anita Lekhwani and Cecilia Tsai, for patiently guiding us through the maze of modern publishing and Saravanan Purushothaman for careful copy-editing that saved us from many errors. Any errors that remain are, of course, our own.

EB, Kingston, Ontario
RAS, Peterborough, Ontario
April 15, 2015

# PREFACE TO THE FIRST EDITION

The role of the solvent in chemical reactions is one of immediate and daily concern to the practicing chemist. Whether in the laboratory or in industry, most reactions are carried out in the liquid phase. In the majority of these, one or two reacting components, or reagents, with or without a catalyst, are dissolved in a suitable medium and the reaction is allowed to take place. The exceptions, some of which are of great industrial importance, are those reactions taking place entirely in the gas phase or at gas–solid interfaces, or entirely in solid phases. Reactions in the absence of solvent are rare, though they include such important examples as bulk polymerization of styrene or methyl methacrylate. Of course, one could argue that the reactants are their own solvent.

Given the importance of solvent, the need for an in-depth understanding of a number of cognate aspects seems obvious. In the past, many texts of inorganic and organic chemistry did not bother to mention that a given reaction takes place in a particular solvent or they mentioned the solvent only in a perfunctory way. Explicit discussion of the effect of changing the solvent was rare, but this is changing. Recent texts, for example, Carey (1996), Clayden *et al.* (2001), Solomons and Fryhle (2000), Streitwieser *et al.* (1992), devote at least a few pages to solvent effects. Morrison and Boyd (1992) and Huheey *et al.* (1993) each devote a whole chapter to the topic.

It is the aim of this monograph to amplify these brief treatments, and so to bring the role of the solvent to the fore at an early stage of the student's career. Chapter 1 begins with a general introduction to solvents and their uses. While it is assumed that the student has taken courses in the essentials of thermodynamics and kinetics, we make no apology for continuing with a brief review of essential aspects of these concepts. The approach throughout is semiquantitative, neither quite elementary nor fully advanced. We have not avoided necessary mathematics, but have made no attempt at rigor, preferring to outline the development of unfamiliar formulas only in sufficient detail to avoid mystification.

The physical properties of solvents are first brought to the fore in Chapter 2, entitled "The Solvent as Medium," which highlights, for example, Hildebrand's solubility parameter, and the Born and Kirkwood–Onsager electrostatic theories. An introduction to empirical parameters is also included. Chapter 3, "The Solvent as Participant," deals chiefly with the ideas of acidity and basicity and the different forms in which they may be expressed. Given the complexities surrounding the subject, the student is introduced in Chapter 4 to empirical correlations of solvent properties. In the absence of complete understanding of solvent behavior, one comes to appreciate the attempts that have been made by statistical analysis (*chemometrics*) to rationalize the subject. A more theoretical approach is made in Chapter 5, but even though this is entitled "Theoretical Calculations," there is in fact no rigorous theory presented. Nevertheless, the interested student may be sufficiently motivated to follow up on this topic. Chapters 6 and 7 deal with some specific examples of solvents: dipolar-aprotic solvents like dimethylformamide and dimethyl sulfoxide and more common acidic/basic solvents, as well as chiral solvents and the currently highlighted room-temperature ionic liquids. The monograph ends with an appendix, containing general tables. These include a table of physical properties of assorted solvents, with some notes on safe handling and disposal of wastes, lists of derived and empirical parameters, and a limited list of values.

A few problems have been provided for some of the chapters.

We were fortunate in being able to consult a number of colleagues and students, including (in alphabetical order) Peter F. Barrett, Natalie M. Cann, Doreen Churchill, Robin A. Cox, Robin Ellis, Errol G. Lewars, Lakshmi Murthy, Igor Svishchev, and Matthew Thompson, who have variously commented on early drafts of the text, helped us find suitable examples and references, helped with computer problems, and corrected some of our worst errors. They all have our thanks.

Lastly, in expressing our acknowledgments we wish to give credit and our thanks to Professor Christian Reichardt, who has written the definitive text in this area with the title *Solvents and Solvent Effects in Organic Chemistry* (2nd Edn., 1988, 534 p.). It has been an inspiration to us to read this text and on many occasions we have been guided by its authoritative and comprehensive treatment. It is our hope that having read our much shorter and more elementary monograph, the student will go to Reichardt's text for deeper insight.

EB, Kingston, Ontario  
RAS, Peterborough, Ontario  
HW, Montreal, Quebec  
February, 2002

# 1

# PHYSICOCHEMICAL FOUNDATIONS

## 1.1 GENERALITIES

The alchemists' adage, "*Corpora non agunt nisi fluida*," "Substances do not react unless fluid," is not strictly accurate, for crystals can be transformed by processes of nucleation and growth. There is growing interest in "mechanochemical" processes, which are carried out by grinding solid reagents together (and which no doubt involve a degree of local melting). Nevertheless, it is still generally true enough to be worthy of attention. Seltzer tablets, for instance, must be dissolved in water before they react to evolve carbon dioxide. The "fluid" state may be gaseous or liquid, and the reaction may be a homogeneous one occurring throughout a single gas or liquid phase, or a heterogeneous one occurring only at an interface between a solid and a fluid, or at the interface between two immiscible fluids. As the title suggests, this book is concerned mainly with homogeneous reactions, and will emphasize reactions of substances dissolved in liquids of various kinds.

The word "solvent" implies that the component of the solution so described is present in excess; one definition is "the component of a solution that is present in the largest amount." In most of what follows it will be assumed that the solution is dilute. We will not attempt to define how dilute is "dilute," except to note that we will routinely use most physicochemical laws in their simplest available forms, and then require that all solute concentrations be low enough that the laws are valid, at least approximately.

Of all solvents, water is of course the cheapest and closest to hand. Because of this alone it will be the solvent of choice for many applications. In fact, it has dominated our thinking for so long that any other solvent tends to be tagged *nonaqueous*, as if water were in some essential way unique. It is true that it has an unusual combination of properties (see, e.g., Marcus, 1998, pp. 230–232). One property in which it is nearly unique is a consequence of its ability to act both as an acid and as a base.

That is the enhanced apparent mobility of the $H_3O^+$ and $HO^-$ ions, explained by the Grotthuss mechanism (Cukierman, 2006; de Grotthuss, 1806):

$$H_3O^+ + H_2O + H_2O \rightarrow H_2O + H_3O^+ + H_2O \rightarrow H_2O + H_2O + H_3O^+ \cdots$$

$$HO^- + H_2O + H_2O \rightarrow H_2O + HO^- + H_2O \rightarrow H_2O + H_2O + HO^- \cdots$$

in which protons hop from one molecule or ion to the next following the electric field, without actual motion of the larger ion through the liquid. This property is shared (in part) with very few solvents, including methanol and liquid hydrogen fluoride, but not liquid ammonia, as may be seen from the ionic equivalent conductances (see Table 1.1). It is apparent that in water, both the positive and negative ions are anomalously mobile. In ammonia neither is, in hydrogen fluoride only the negative ion is, and in methanol only the positive ion is.

As aqueous solution of an acid is diluted by addition of a solvent that does not contribute to the hydrogen-bonded network, the Grotthuss mechanism becomes less effective. For an electrolyte that conducts electricity by migration of ordinary ions through the solvent, Walden observed that the product of the limiting equivalent conductance of the electrolyte with the viscosity in different solvent or mixtures of different composition is approximately constant. The limiting equivalent conductance of HCl in several dioxane/water mixtures was measured by Owen and Waters (1938). As can be seen in Figure 1.1, in 82% dioxane the Walden product drops to hardly a quarter of its maximum. The Grotthuss mechanism is largely suppressed.

More and more, however, other solvents are coming into use in the laboratory and in industry. Aside from organic solvents such as alcohols, acetone, and hydrocarbons, which have been in use for many years, industrial processes use such solvents as sulfuric acid, hydrogen fluoride, ammonia, molten sodium hexafluoroaluminate (cryolite), various other "ionic liquids" (Welton, 1999), and liquid metals. Jander and Lafrenz (1970) cite the industrial use of bromine to separate caesium bromide (sol'y 19.3 g/100 g bromine) from the much less soluble rubidium salt. The list of solvents available for preparative and analytical purposes in the laboratory now is long and growing, and though water will still be the first solvent that comes to mind, there is no reason to stop there.

**TABLE 1.1   Limiting Equivalent Conductances of Ions in Amphiprotic Solvents**

| In $H_2O$ at 25°C | | In $NH_3$ at −33.5°C[a] | | In HF at 20°C[b] | | In MeOH at 25°C[c] | |
|---|---|---|---|---|---|---|---|
| $H_3O^+$ | 349.8 | $NH_4^+$ | 131 | $H_2F^+$ | 102 | $MeOH_2^+$ | 141.8 |
| $HO^-$ | 198.5 | $NH_2^-$ | 133 | $HF_2^-$ | 350 | $MeO^-$ | 53.02 |
| $Na^+$ | 50.11 | $Na^+$ | 130 | $Na^+$ | 150 | $Na^+$ | 45.5 |
| $K^+$ | 73.52 | $K^+$ | 168 | $K^+$ | 150 | $K^+$ | 53.6 |

[a] Kraus and Brey (1913).
[b] Kilpatrick and Lewis (1956).
[c] Ogston (1936), Conway (1952, pp. 155, 162).

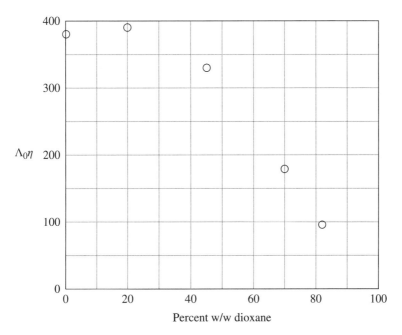

**FIGURE 1.1** The Walden product, $\Lambda_0\eta$, for HCl in 1,4-dioxane/water mixtures versus percentage of dioxane at 25°C. Data from Owen and Waters (1938).

After the first observation of the effect of solvent change on reaction rate by Berthelot and Pean de St. Gilles (1862) and the first systematic study, Menschutkin (1887, 1890), the study of solvent effects was for some years largely the work of physical–organic chemists. The pioneer in this growing field was Hammett and Deyrup (1932, and see his book, *Physical Organic Chemistry*, 1970). The study of solvent effects was pursued notably by Hughes and Ingold (1935) and Grunwald and Winstein (1948). One of us (R. A. S.) was privileged to attend Ingold's lectures at Cornell that became the basis of his book (Ingold, 1969), while E. B. can still recall vividly the undergraduate lectures by both Hughes and Ingold on the effect of solvent in nucleophilic substitution: the Hughes–Ingold Rules (Ingold, 1969). Inorganic chemists soon followed. Tobe and Burgess (1999, p. 335) remark that while inorganic substitution reactions of known mechanism were used to probe solvation and the effects of solvent structure, medium effects have been important in understanding the mechanisms of electron transfer.

If a solvent is to be chosen for the purpose of preparation of a pure substance by synthesis, clearly the solvent must be one that will not destroy the desired product, or transform it in any undesirable way. Usually it is obvious what must be avoided. For instance, one would not expect to be able to prepare a strictly anhydrous salt using water as the reaction medium. Anhydrous chromium (III) chloride must be prepared by some reaction that involves no water at all, neither in a solvent mixture nor in any of the starting materials, nor as a by-product of reaction. A method that works uses the reaction at high temperature of chromium (III) oxide with tetrachloromethane (carbon tetrachloride), according to the equation:

$$Cr_2O_3(s) + 3CCl_4(g) \rightarrow 2CrCl_3(s) + 3COCl_2(g)$$

Here no solvent is used at all.[1] Some other anhydrous salts may be prepared using such solvents as sulfur dioxide, dry diethyl ether (a familiar example is the Grignard reaction, in which mixed halide–organic salts of magnesium are prepared as intermediates in organic syntheses), hydrogen fluoride, and so on.

A more subtle problem is to maximize the yield of a reaction that could be carried out in any of a number of media. Should a reaction be done in a solvent in which the desired product is most or least soluble, for instance? The answer is not immediately clear. In fact one must say, "It depends…." If the reaction is between ions of two soluble salts, the product will precipitate out of solution if it is insoluble. For example, a reaction mixture containing barium, silver, chloride, and nitrate ions will precipitate insoluble silver chloride if the solvent is water, but in liquid ammonia the precipitate is barium chloride. Another example, from organic chemistry, described by Collard *et al.* (2001) as an experiment suitable for an undergraduate laboratory, is the dehydrative condensation of benzaldehyde with pentaerythritol in aqueous acid to yield the cyclic acetal, 5,5-bis(hydroxymethyl)-2-phenyl-1,3-dioxane, **1**:

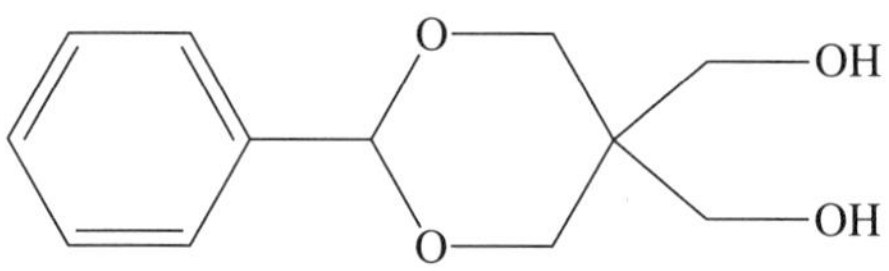

**1**

At 30°C the product is sufficiently insoluble to appear as a precipitate, so the reaction proceeds in spite of the formation of water as by-product. On the other hand, we will show in Chapter 2 that, in a situation where all the substances involved in a reaction among molecules are more or less soluble, *the most soluble substances will be favored at equilibrium.*

## 1.2 CLASSIFICATION OF SOLVENTS

Solvents may be classified according to their physical and chemical properties at several levels. The most striking differences among liquids that could be used as solvents are observed between *molecular liquids, ionic liquids* (molten salts or salt mixtures, room-temperature ionic liquids), and *metals.* They can be considered as extreme types, and represented as the three vertices of a triangle (Trémillon, 1974) (see Fig. 1.2). Intermediate types or mixtures can then be located along edges or within the triangle. The room-temperature ionic liquids (see later, Section 8.3), which

[1]Caution: The reagent tetrachloromethane and the by-product phosgene are toxic and environmentally undesirable.

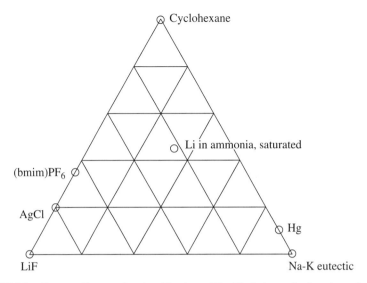

**FIGURE 1.2** Ternary diagram for classification of liquids (schematic; location of points is conjectural); [bmim]PF$_6$ represents a room-temperature ionic liquid (see Section 8.3). After Trémillon (1974).

typically have large organic cations and fairly large anions, lie along the molecular–ionic edge, for instance.

Among the molecular liquids, further division based on physical and chemical properties leads to categories variously described (Barthel and Gores, 1994; Reichardt and Welton, 2011) as **inert** (unreactive, with low or zero dipole moments and low polarizability), **inert-polarizable** (e.g., aromatics, polyhalogenated hydrocarbons), **protogenic** (hydrogen-bonding proton donors, HBD), **protophilic** (hydrogen-bonding proton acceptors, HBA), **amphiprotic** (having both HBD and HBA capabilities), and **dipolar-aprotic** (having no marked HBD or HBA tendencies, but possessing substantial dipole moments). Examples of these classes are listed in Table 1.2. The ability of solvent molecules to act as donors or acceptors of electron pairs, that is, as Lewis bases or acids, complicates the classification. Nitriles, ethers, dialkyl sulfides, and ketones are electron-pair donors (EPD), for example; sulfur dioxide and tetracyanoethene are electron-pair acceptors (EPA). EPD and EPA solvents can be further classified as *soft* or *hard*. (Classifying can be habit-forming.) Pushing the conditions can cause normally inert substances to show weak prototropic properties: dimethyl sulfoxide can lose a proton to form the *dimsyl* ion, $CH_3SOCH_2^-$, in very strongly basic media (Olah *et al.*, 1985). An equilibrium concentration of dimsyl ion, very small, though sufficient for hydrogen–deuterium isotopic exchange to occur between dimethyl sulfoxide and $D_2O$, is set up even in very dilute aqueous NaOH (Buncel *et al.*, 1965). Carbon monoxide, not normally considered a Brønsted base, can be protonated in the very strongly acidic medium of HF–SbF$_5$ (de Rege *et al.*, 1997).

**TABLE 1.2   Molecular Solvents**

| Classes | Examples |
|---|---|
| Inert | Aliphatic hydrocarbons, fluorocarbons |
| Inert-polarizable | Benzene ($\pi$-EPD), tetrachloromethane, carbon disulfide, tetracyanoethene ($\pi$-EPA) |
| Protogenic (HBD) | Trichloromethane |
| Protophilic (HBA) | Tertiary amines (EPD) |
| Amphiprotic | Water, alcohols; ammonia is more protophilic than protogenic, while hydrogen fluoride is the reverse |
| Dipolar-aprotic | Dimethylformamide, acetonitrile (EPD, weak HBA), dimethyl sulfoxide, hexamethylphosphortriamide |

## 1.3   SOLVENTS IN THE WORKPLACE AND THE ENVIRONMENT

The majority of solvents must be considered as toxic to some degree. Quite aside from those that have specific toxicity, whether through immediate, acute effects, or more insidiously as, for instance, carcinogens the effects of which may take years to manifest, all organic substances that are liquid at ordinary temperatures and are lipophilic (fat-soluble) are somewhat narcotic. The precautions that should be taken depend very much on their individual properties. Inhalation of vapors should always be avoided as much as possible. Many solvents are quickly absorbed through the skin. Use of an efficient fume hood is always advisable. Protective gloves, clothing, masks, and so on, should be available and used as advised by the pertinent literature (in Canada, the Material Safety Data Sheet). The rare solvents that exhibit extreme toxicity, such a liquid HCN or HF, require special precautions. The latter is an example of substances absorbed rapidly through the skin, with resulting severe burns and necrosis. Most common solvents are inflammable to varying degrees.[2] Those with low boiling points or low flash points (see Table A.1) require special precautions. A few have in addition particularly low ignition temperatures; a notable example is carbon disulfide, the vapor of which can be ignited by a hot surface, without a flame or spark. Transfer of a solvent with low electrical conductivity from a large shipping container to a smaller, ready-use container can be associated with an accumulation of static charge, with the chance that a spark may occur, causing fire. Proper grounding of both containers can prevent this.

Environmental concerns include toxicity to organisms of all sorts, but perhaps more importantly the tendency of each substance to persist and to be transported over long distances. Chemical stability may seem to be a desirable property, but unless a solvent is biodegradable or easily decomposed photochemically by sunlight, it can become a long-lasting contaminant of air, water, or soil, with consequences that we

---

[2]In 1978, the Canadian Transportation of Dangerous Goods Code was modified to require that labels on goods that burn easily are to use the word "inflammable" only (Johnstone, 1978; Stairs, 1978b).

probably cannot foresee. Much effort is currently going into the consideration of the long-term effects of industrial chemicals, including solvents, should they escape.

For these reasons, selection of a solvent should always be made with an eye on the effects it might have if it is not kept to minimum quantities and recycled as much as possible. Consideration should also be given to the history of the solvent before it reaches the laboratory. Does its manufacture involve processes that pose a danger to the workers or to the environment? These matters are discussed further in Section 8.6.

## 1.4 SOME ESSENTIAL THERMODYNAMICS AND KINETICS: TENDENCY AND RATE

How a particular reaction goes or does not go in given circumstances depends on two factors, which may be likened, "psychochemically" speaking, to "wishing" and "being able."[3] The first is the **tendency to proceed**, or the degree to which the reaction is out of equilibrium, and is related to the equilibrium constant and to free energy changes (Gibbs or Helmholtz). It is the subject of **chemical thermodynamics**. The second is the speed or **rate** at which the reaction goes, and is discussed in terms of rate laws, mechanisms, activation energies, and so on. It is the subject of **chemical kinetics**. We will need to examine reactions from both points of view, so the remainder of this chapter will be devoted to reviewing the essentials of these two disciplines, as far as they are relevant to our needs. The reader may wish to consult, for example, Atkins and de Paula (2010), for fuller discussions of relevant thermodynamics and kinetics.

## 1.5 EQUILIBRIUM CONSIDERATIONS

For a system at constant pressure, which is the usual situation in the laboratory when we are working with solutions in open beakers or flasks, the simplest formulas to describe equilibrium are written in terms of the Gibbs energy $G$, and the enthalpy $H$. For a reaction having an equilibrium constant $K$ at the temperature $T$, one may write:

$$\Delta G_T^0 = -RT \ln K \qquad (1.1)$$

$$\Delta H_T^0 = -R \left( \frac{\partial \ln K}{\partial (1/T)} \right)_P \qquad (1.2)$$

The equilibrium constant $K$ is of course a function of the activities of the reactants and products, for example, for a reaction: $A + B \rightleftharpoons Y$

---

[3]There is a word, very pleasing to us procrastinators, "velleity," which is defined (Fowler *et al.*, 1976) as "low degree of volition not prompting to action." See also Nash (1938).

$$K = \frac{a_Y}{a_A \cdot a_B} \tag{1.3}$$

By choice of standard states one may express the activities on different scales. For reactions in the gas phase, it is convenient, and therefore common, to choose a standard state of unit activity on a scale of pressure such that the limit of the value of the dimensionless activity coefficient, $\gamma = a_i/P_i$, as the pressure becomes very low, is unity. The activity on this scale is expressed in pressure units, usually atmospheres or bars, so we may write

$$K = \frac{\gamma_Y P_Y}{\gamma_A P_A \gamma_B P_B} = \Gamma_\gamma K_P \tag{1.4}$$

The activity coefficient quotient $\Gamma_\gamma$ is unity for systems involving only ideal gases, and for real gases at low pressure.

For reactions involving only condensed phases, including those occurring in liquid solutions, which are our chief concern, the situation is very different. Three choices of standard state are in common use. For the solvent (i.e., the substance present in largest amount), the standard state almost universally chosen is the pure liquid. This choice is also often made for other liquid substances that are totally or largely miscible with the solvent. The activity scale is then related to the mole fraction, through the *rational activity coefficient f*, which is unity for each pure substance. For other solutes, especially those that are solid when pure, or for ionic species in solution in a nonionic liquid, activity scales are used that are related either to the molar concentration or the molality, depending on experimental convenience. On these scales, the activity coefficients become unity in the limit of low concentration.

If a substance present in solution is to some extent volatile, that is, if it exerts a measurable vapor pressure, its activity in solution can be related to its activity in the gas (vapor) phase. If the solution is ideal, all components obey Raoult's Law, expressed by Equation 1.5, and illustrated by the dashed lines in Figure 1.3.

$$p_i = p_i^0 x_i \tag{1.5}$$

Here $p_i$ is the vapor pressure of the $i$th substance over the solution, $p_i^0$ is the vapor pressure it would exert in its standard (pure liquid) state, and $x_i$ is its mole fraction in the solution. We can now define an "absolute" activity (not really absolute, but relative to the gas phase standard state on the pressure scale as earlier) measured by $p_i$, assuming that the vapor may be treated as an ideal gas or by the *fugacity*[4] if necessary. We shall always make the *"ideal gas" assumption*, without restating it.

---

[4]Fugacity *f* is pressure corrected for nonideality. It is defined so that the Gibbs energy change on isothermal, reversible expansion of a mole of a real gas is $\Delta G = \int V dP = RT \ln(f/f_0)$. For a real gas at low enough pressures, $f = P$. Fugacities can be calculated from the equation of state of the gas if needed. See any physical chemistry textbook, for example, Atkins and de Paula (2010, pp. 129–130). For an only slightly nonideal gas $f = P^2 V_m/RT$, approximately.

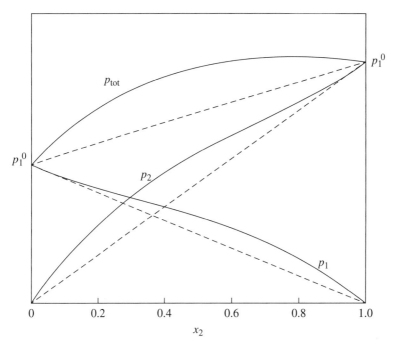

**FIGURE 1.3** Vapor pressure over binary solutions. Dashed lines: ideal (Raoult's Law). Solid curves: positive deviations from Raoult's Law. Note that where $x_2 \ll 1$, $P_1$ is close to ideal, and vice versa.

## 1.6 THERMODYNAMIC TRANSFER FUNCTIONS

The thermodynamic equilibrium constant as defined earlier is independent of the solvent. The practical equilibrium constant is not, because the activity coefficients of the various reactant and product species will change in different ways when the reaction is transferred from one solvent to another. One way of considering these changes is through the use of thermodynamic transfer functions. The standard Gibbs energy of a reaction in a solvent $\mathbf{S}$, $\Delta G_{\mathbf{S}}^0$, may be related to that in a reference solvent $\mathbf{O}$, $\Delta G_{\mathbf{O}}^0$, by considering the change in Gibbs energy on transferring each reactant and product species from the reference solvent to $\mathbf{S}$. The reference solvent may be water or the gas phase (no solvent). Other functions (enthalpy, entropy) can be treated in the same fashion as $G$. A reaction converting reactants $\mathbf{R}$ to products $\mathbf{P}$ in the two solvents can be represented in a Born–Haber cycle:

$$(\text{In } \mathbf{S}) \, R(S) \xrightarrow{\Delta G_{\mathbf{S}}^0} P(S)$$
$$\delta_{\text{tr}} G^{(R)} \quad - \qquad\qquad - \delta_{\text{tr}} G^{(P)}$$
$$(\text{In}\mathbf{O}) \, R(0) \xrightarrow{\Delta G_{\mathbf{O}}^0} P(0)$$

$$\Delta G_S^0 = -\delta_{tr} G^{(R)} + \Delta G_O^0 + \delta_{tr} G^{(P)} \tag{1.6}$$

For each participating substance **I**, the term $\delta_{tr} G^{(I)}$ can be obtained from vapor pressure, solubility, electrical potential, or other measurements that enable the calculation of activity coefficients and hence of standard Gibbs energies, using Equation 1.7.

$$\delta G_{tr}^{(i)} = G_S^{0(i)} - G_O^{0(i)} \tag{1.7}$$

Since the Gibbs energy and the activity coefficient are related through Equation 1.8, this development could have been carried out in terms of $\ln a$ or $\ln f$.

$$G_S^{0(i)} - G_O^{0(i)} = RT \ln \left( \frac{a_S^{0(i)}}{\alpha_O^{0(i)}} \right) = RT \ln \left( \frac{f_S^i}{f_O^i} \right) \tag{1.8}$$

Because of the analogy between the transition states in kinetics and the products in equilibrium (see later, Section 1.6), similar considerations can be applied to the understanding of solvent effects on reaction rates. This will be illustrated in Chapter 6.

## 1.7   KINETIC CONSIDERATIONS: COLLISION THEORY

Elementary reactions occurring in the gas phase have been fruitfully discussed in terms derived from the Kinetic–Molecular Theory of Gases. The result is Equation 1.9,

$$\text{Rate} = PZ_0 [A][B] e^{-E_a/PT} \tag{1.9}$$

$$Z_0 = \pi d^2 \left( \frac{8k_B T}{\pi \mu} \right)^{1/2} N^2 \tag{1.10}$$

where $Z_0$ is the number of collisions per unit time between A and B molecules at unit concentrations given by Equation 1.10, [A] and [B] represent the concentrations of the reacting species, $d$ is the mean diameter of A and B, $k_B$ is the Boltzmann constant, and $\mu$ their reduced mass, and $E_a$ is the activation energy. $P$ is the steric or probability factor, that is, the probability that the colliding molecules are in suitable orientations and internal configuration to permit reaction, as illustrated in Figure 1.4. The factors $PZ_0$ are usually combined to form the *Arrhenius pre-exponential factor*, usually denoted by $A$. Equations 1.9 and 1.10 have allowed a substantial level of understanding of simple reactions to be achieved, and by combining elementary steps into multistep mechanisms, complex reactions may also be described. This simple Arrhenius treatment is not applicable to reaction in solution, however, so for our purposes another approach is needed.

(a)

(b)

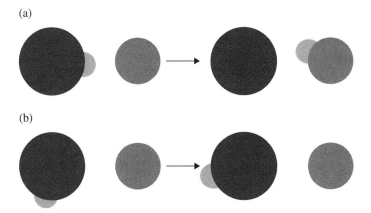

**FIGURE 1.4** Successful (a) and unsuccessful (b) transfer of a hydrogen atom from HI to Cl.

## 1.8 TRANSITION-STATE THEORY

The variously named *transition-state theory* (the preferred name) or *absolute reaction rate theory*, developed by Eyring and associates (Berry *et al.*, 2000, pp. 911–927; Eyring, 1935; Laidler and Meiser, 1995, pp. 382–387) and by Evans and Polanyi (1935), takes a quite different view. The reacting molecules are considered as entering a "transition state," forming an "activated complex," which resembles an ordinary molecule in all respects but one, which is that one of its normal modes of vibration is *not* a vibration, because there is no restoring force; rather it will lead to decomposition of the complex, either to form the products of the reaction or to reform the starting molecules. Quantum–mechanical calculations of the energetics and geometry of molecules in configurations that represent transition states can be carried out using such computer programs as GAUSSIAN, SPARTAN, or HYPERCHEM (Levine, 2013). Of the normal modes of vibration of such a transition-state "molecule," one has a *negative force constant*. What is meant by this is that there is no force restoring the molecule to an equilibrium configuration in the direction of this motion; in fact the force is repulsive, leading to rearrangement or decomposition, to form the products of the reaction, or to reform the starting molecules. Since the force constant is negative, the frequency, which depends on the square root of the force constant, contains the factor $\sqrt{-1}$; that is, it is imaginary. A graph of the energy of the system as a function of the normal coordinates of the atoms (the *potential energy surface*) in the vicinity of the transition state takes the form of a *saddle* or *col*, illustrated in Figure 1.5.

From the saddle point, the energy increases in all the principal directions except along the direction that leads to reaction (forward) or (backward) to reform the starting materials. The course of a simple reaction may be represented as motion along the *reaction coordinate*, which is a combination of atomic coordinates leading from the initial configuration (reactants) through the transition state to the final

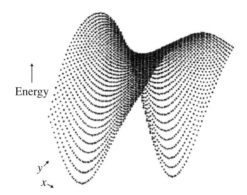

Energy

$y$

$x$

**FIGURE 1.5**    A portion of a potential-energy surface $E(x,y)$, showing a saddle point.

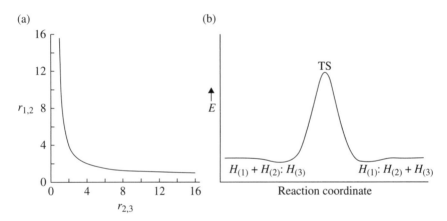

(a)                                                          (b)

16

12

$r_{1,2}$  8                                        $E$

4

0

0      4      8      12      16

$r_{2,3}$

TS

$H_{(1)} + H_{(2)}: H_{(3)}$              $H_{(1)}: H_{(2)} + H_{(3)}$

Reaction coordinate

**FIGURE 1.6**    (a) The reaction pathway of least energy and (b) the profile along the pathway, for the hydrogen atom–molecule exchange reaction (schematic).

configuration (products) along, or nearly along, the path of least energy. Figure 1.6a shows a projection of the path of least energy on the potential energy surface for a very simple reaction, in which a hydrogen atom attacks a hydrogen molecule directly at one end, and one atom is transferred to the attacking atom. The reaction coordinate is measured along the (approximately hyperbolic) pathway. The energy as a function of the reaction coordinate is shown in Figure 1.6b.

In most reactions, especially those taking place in solution, the situation is more complicated. For instance, Figure 1.7 shows a possible form of the energy profile for a reaction in which one ligand in a transition-metal complex ion is replaced by another. In the scheme here, M represents a trivalent metal ion. There may first be formed an outer-sphere complex, perhaps an ion pair (1.11), which then rearranges (1.12) so the arriving and leaving ligands change places (not necessarily with retention of configuration). The leaving ligand, now in the outer sphere, finally leaves (1.13).

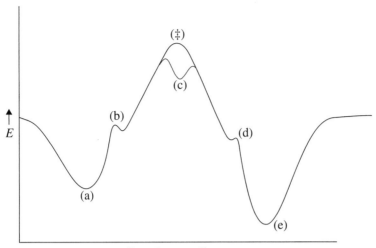

FIGURE 1.7 A possible, more realistic reaction profile for a ligand-exchange reaction, showing reactants (a), precursor (b), and successor (d) complexes, the transition state ($\ddagger$), the possibility of the formation of a reactive intermediate (c), and products (e). Redrawn after Kettle (1996) by permission of Oxford University Press.

$$X^- + M(H_2O)_5 Y^{2+} \rightleftharpoons X^- M^{(III)}(H_2O)_6 Y^{2+} \tag{1.11}$$

$$X^- M(H_2O)_6 Y^{2+} \rightarrow [M^{(III)}(H_2O)_6 XY]^+ \rightarrow Y^- M(H_2O)_6 X^{2+} \tag{1.12}$$

(The seven-coordinate species may be a true activated complex, corresponding to the curve with a single maximum in Figure 1.7, or a transient intermediate, corresponding to the light curve with a dip between two maxima.)

$$Y^- M(H_2O)_6 X^{2+} \rightarrow M(H_2O)_6 X^{2+} + Y^- \tag{1.13}$$

The composition of the activated complex may be deduced from the rate law. In the case of a multistep reaction, if one step is rate controlling, which is usually true, the composition of the activated complex of the rate-controlling step may still be deduced from the rate law for the overall reaction. For example, if a reaction between two substances A and B follows a rate law of the form of Equation 1.14 (over certain ranges of concentrations and temperatures):

$$\text{Rate} = k[A]^n[B]^m \tag{1.14}$$

the activated complex has a composition represented by $A_n B_m$. (There are some subtle aspects of this rule; e.g., see Problem 2.2.) We know nothing, however, of its structure nor of the steps in the reaction, in the absence of other evidence.

Nevertheless, one may write a statement resembling an equilibrium-constant expression, relating the activities or, approximately, the concentrations of the reactants and the activated complex (represented by the "double-dagger" or "Cross of Lorraine" symbol ‡):

$$K^{\ddagger} = \frac{[\ddagger]}{[A]^n [B]^m} \tag{1.15}$$

Then, if we assume that the rate of decomposition of the complex is first order, that is, that it reacts to form the products at a rate proportional to its concentration, we obtain:

$$\text{Rate} \propto [\ddagger] \propto [A]^v [B]^\mu \tag{1.16}$$

That is, the ordinary rate constant, $k$, is proportional to $K^{\ddagger}$.

Specifically, one may write (Atkins and de Paula, 2010, p. 846; Laidler and Meiser 1995, p. 741):

$$k = \kappa \frac{k_B T}{h} K^{\ddagger} \tag{1.17}$$

where $k_B$ is the Boltzmann's constant, $h$ is the Planck's constant, and $T$ is the absolute temperature. The group $k_B T/h$ has a value of about $6\,\text{ps}^{-1}$ at 298 K. The factor $\kappa$ is a constant, the *transmission coefficient*, the value of which is close to unity for bimolecular reactions in the gas phase. Abboud *et al.* (1993, p. 75) cite a computer simulation study (Wilson, 1989) of the chloride exchange reaction:

$$*Cl^- + H_3C - Cl \rightarrow *Cl - CH_3 + Cl^-$$

in which the transmission coefficient was calculated to be unity for the reaction in the gas phase, but 0.55 in aqueous solution, apparently owing to confinement of the reacting species within a "cage" of water molecules, so that multiple crossings of the transition barrier can occur.

To the extent that $K^{\ddagger}$ can be considered an ordinary equilibrium constant, one may then apply the usual thermodynamic relations, that is,

$$\Delta_{\ddagger} G^0 = -RT \ln K^{\ddagger} \tag{1.18}$$

$$\Delta_{\ddagger} H^0 = -R \left( \frac{\partial \ln K^{\ddagger}}{\partial (1/T)} \right)_P \tag{1.19}$$

$$\Delta_{\ddagger}S^0 = \frac{\Delta_{\ddagger}H^0 - \Delta_{\ddagger}G^0}{T} \tag{1.20}$$

Equation 1.17 may then be written:

$$k = \kappa \frac{k_BT}{h} e^{-\Delta_{\ddagger}G^0/RT} = \kappa \frac{k_BT}{h} e^{\Delta_{\ddagger}S^0/R} e^{-\Delta_{\ddagger}H^0/RT} \tag{1.21}$$

or, in logarithmic form,

$$\ln k = \ln\left(\frac{\kappa k_B}{h}\right) + \frac{\Delta_{\ddagger}S^0}{R} + \ln T - \frac{\Delta_{\ddagger}H^0}{RT} \tag{1.22}$$

and differentiating Equation 1.22 with respect to $(1/T)$,

$$\frac{d(\ln k)}{d(1/T)} = -T - \frac{\Delta_{\ddagger}H^0}{T} = -\frac{\Delta_{\ddagger}H^0 + RT}{R} = -\frac{E_a}{R} \tag{1.23}$$

where $E_a$ is the ordinary (Arrhenius) experimental activation energy, which is thus equal to $\Delta_{\ddagger}H^0 + RT$.

The two theoretical approaches, one in terms of molecular collisions and the other in terms of an activated complex, are not opposed, but complementary. A key to the connection between them is the entropy of activation. When both the rate constant and the temperature coefficient of the rate constant are known, $\Delta_{\ddagger}G^0$ and $\Delta_{\ddagger}H^0$ ($=E_a - RT$) can be used with Equation 1.20 to obtain $\Delta_{\ddagger}S^0$. In an ordinary bimolecular reaction with no special steric requirements, the formation of the activated complex means the formation of one rather "loose" molecule from two. A negative entropy change is to be expected, perhaps comparable to that for the combination of two iodine atoms (Atkins and de Paula, 2010, p. 922),

$$2I(g) \rightarrow I_2(g) : \Delta S_{298}^0 = -100.9 \text{ J mol}^{-1}.$$

A value much more negative than this implies the loss of much freedom of motion on formation of the complex, and corresponds to a small value of $P$, the steric factor in the collision theory. On the other hand, less negative or even positive values of $\Delta_{\ddagger}S^0$ occasionally occur, though rarely if ever for bimolecular reactions in the gas phase. They imply that the complex is very loosely bound, or, in solution, that the complex is less tightly solvated than are the reactant species.

Most reactions in the gas phase at low pressures can be treated as if no foreign molecules (i.e., other than reactants, intermediates, or products of the reaction) are present. Thus the presence of an inert gas such as argon is not important. An exception to this rule is any reaction in which two atoms combine to form a stable diatomic molecule. This cannot happen unless some means exists of getting rid of the energy of formation of the bond. A *third body*, which may be any molecule or the container wall, must be present to absorb some of this energy. Its function has been likened (more "psychochemistry?") to that of a chaperon (Laidler, 1987, p. 183, after G. Porter), present not to prevent union but to ensure that the union is stable and is not formed in an excited state. An extensive literature exists on the efficacy of different molecules as third bodies (Mitchell, 1992; Troe, 1978), and on the influence of the container wall in this and other ways.

## 1.9  REACTIONS IN SOLUTION

When it comes to reactions in solution, the results of kinetic experiments are difficult to understand, except qualitatively, through the collision theory. The very concept of a collision is hard to define in the liquid phase, in which molecules are not free to travel in straight lines between collisions, but move in constant interaction with neighbors, in a "tipsy reel" (J. H. Hildebrand's phrase). What happens when two solute molecules come into contact in solution, perhaps to react, or perhaps to diffuse apart unchanged, is sometimes called an "encounter," rather than a collision. Rabinowitch and Wood (1936) demonstrated this by the use of a model in which a few metal balls rolled about on a level table, making collisions that were detected electrically. When many nonconducting balls were added to the set on the table, so that it became rather crowded, instead of single collisions at long and irregular intervals, collisions happened in groups, while the two metal balls were temporarily trapped in a cage of other balls. Computer modeling in three dimensions, using simulated hard spheres, gave a similar result: collisions in a crowded space between labeled molecules occurred in groups of 10 to nearly 100, depending on the degree of crowding. In the hard-sphere representation, collisions could still be recognized. In a more realistic computer model in which molecular attractions and repulsions are both dependent on distance (which enormously increases the amount of calculation required), an encounter would become a continuous interaction of a complicated kind. During the encounter, something resembling a definite complex, called an *encounter complex*, is present (Langford and Tong, 1977). Eigen and Tamm (1962), in work on ultrasonic effects on solutions of sulfates of divalent metals, interpreted their data as showing that such an encounter complex was formed between the oppositely charged ions, but an encounter complex may exist in the absence of such electrostatic assistance.

## 1.10  DIFFUSION-CONTROLLED REACTIONS

Consider a bimolecular reaction in solution as occurring in two steps. In the first step, an encounter complex is formed:

$$A + B \rightarrow [AB] \quad k_1$$

The complex may then either revert to separated reactants or react to form products:

$$[AB] \rightarrow A + B \quad k_{-1}$$

$$[AB] \rightarrow P \quad k_2$$

Applying the steady-state assumption to the concentration of the encounter complex:

$$\frac{d[AB]}{dt} = k_1[A][B] - k_{-1}[AB] - k_2[AB] = 0 \tag{1.24}$$

it may be shown that the rate of formation of products is given by Equation 1.29:

$$\frac{d[P]}{dt} = k_2[AB] = k[A][B]; \ k = \frac{k_2 k_1}{k_2 + k_{-1}} \tag{1.25}$$

If the encounter complex reacts to form products much faster than it reverts to reactants, that is, if $k_2 \gg k_{-1}$, then $k = k_1 k_2 / k_2 = k_1$, that is, the rate is controlled by the rate of formation of the encounter complex. Such a reaction is described as *diffusion controlled* or *encounter controlled*.

The magnitude of $k_1$ is approximately given by Equation 1.26 (Atkins and de Paula, 2010, p. 839–842; Cox, 1994, p. 59):

$$k_1 \approx \frac{8000RT}{3\eta} \tag{1.26}$$

where $R$ is the gas constant and $\eta$ is the viscosity. A factor of 1000 lets the result be in the conventional units, $1\,mol^{-1}\,s^{-1}$. A reaction, the rate of which is dependent on bond making or breaking when run in an ordinary solvent, may be diffusion controlled when run in such a highly viscous solvent as glycerol (1,2,3-propanetriol). This has been demonstrated with reactions as diverse as solvent exchange in complexes of $Cr^{2+}$, $Cu^{2+}$, and $Ni^{2+}$ (Caldin and Grant, 1973) and reactions of ferroprotoporphyrin IX (2) with CO and with $O_2$ (Caldin and Hasinoff, 1975).

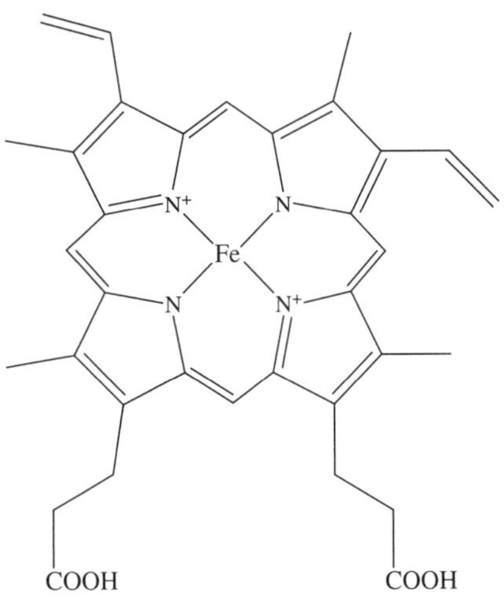

2

Electron-transfer reactions are a class in which diffusion control may be observed. If an electron-donor species D (reductant) is to react with an electron acceptor A (oxidant), they first form an encounter complex, within which the transfer occurs by tunneling at a rate chiefly determined by the height of the barrier between the donor's HOMO and the acceptor's LUMO, and its width, which is determined by the distance of closest approach. If the transfer rate constant is large, the rate-limiting step will be the formation of the encounter complex by diffusion. This picture is an oversimplification. The theory developed by R. A. Marcus, and independently by others, is described in most physical chemistry texts (e.g., Atkins and de Paula, 2010, pp. 856–861).

## 1.11   REACTION IN SOLUTION AND THE TRANSITION-STATE THEORY

The most satisfactory way to consider reactions in solution is through the thermodynamic interpretation of the Transition-State Theory, by examining the effects of various properties of the solvent on the activity of each reactant species and on the activated complex, treating the latter almost as "just another molecule." The solvent can influence the solute molecules by acting on them with "physical" forces (van der Waals forces and electrostatic forces due to the polarity and polarizability of solvent and solute molecules), but also in more obviously "chemical" ways, through the formation of hydrogen bonds or molecular or ionic complexes of various kinds. Changing from an "inert" solvent, one that solvates solutes weakly, to one that exerts stronger forces, may either retard or accelerate a reaction through the change in enthalpy of activation. This depends on whether the latter solvent interacts more strongly with the reactants or with the activated complex. Dewar (1992) discusses an example of a reaction in which the necessity of desolvation of an attacking ion has a profound effect. (Incidentally (p. 160), he describes the hard–soft acid–base distinction as "mythical," at least as an explanation of the difference between nucleophilic substitutions at carbonyl and saturated carbon atoms.) To take the simplest case, if the reactant in a unimolecular reaction (perhaps an isomerization or an $S_N1$ substitution) is more strongly solvated, the reaction will be retarded, through the increase in activation enthalpy; if it is the activated complex that is more solvated, the reverse effect will be found (see Fig. 1.8).

The free energy of solvation is a composite of the enthalpy and entropy. Entropy of solvation can also have large effects. Strong solvation usually implies loss of entropy, owing to relative immobilization of solvent molecules. Strong solvation of the reactant, therefore, makes the entropy of activation more positive, thus (from Eq. 1.20) making the Gibbs energy of activation less positive, and the reaction therefore faster. The effect of strong solvation on the entropy of the activated complex, on the other hand, retards the reaction. Thus, the enthalpy and entropy of solvation of either the reactant or the activated complex have opposite effects. Prediction of the overall effect requires that these be disentangled. The required information concerning reactants is in principle available. That for activated complexes is not,

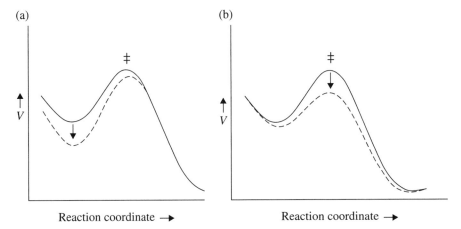

**FIGURE 1.8** Effect of solvation on activation energy. Potential energy $V$ versus the reaction coordinate. Solid curves represent the energy profile in the absence of solvation. (a) Solvation of the reactant (increased activation energy). (b) Solvation of the activated complex (reduced activation energy).

though estimates may be made if data on molecules that resemble a postulated activated complex are known. In favorable cases, the dissection of the kinetics into these parts can be done (Blandamer, 1977; Buncel and Symons, 1981; Buncel and Wilson, 1977, 1979; Tobe and Burgess, 1999, p. 346, 363).

$$R_3N + RI \rightarrow R_3N^+ + I^- \tag{1.27}$$

The Menschutkin Reaction (1.27) provides an example of a case where in polar solvents the solvation of the activated complex has a major effect on the rate. Hartmann and Schmidt (1969) showed that in a series of 12 solvents of increasing polarity from 1,1,1-trichloroethane ($\varepsilon_r = 7.52$, $E_r^N = 0.170$) to nitrobenzene ($\varepsilon_r = 34.78$, $E_r^N = 0.324$) the rate increases by a factor of 52, illustrating the acceleration due to solvation of the activated complex. It also illustrates the importance of the entropy of activation. Here the change in the Gibbs energy of activation is made up of reinforcing contributions from enthalpy and entropy changes; the contribution of the latter at 50°C is 2.5 times that of the former.

The overall effect on the reaction rate thus depends on the free energies of the initial and transition states. The various possibilities, in terms of the free energy, are summarized qualitatively in Table 1.3. Reinforcement occurs if the transfer free energies of reactants and transition state have opposite signs. If they have the same sign, partial or complete balancing is expected.

For reactions in solution an additional thermodynamic property that can be helpful is available. The effect of pressure on the equilibrium constant of a reaction yields the volume change of reaction, $\Delta V$, given by Equation 1.28.

**TABLE 1.3   Transfer Free Energies of Reactants ($\delta_{tr}G^R$) and Transition States ($\delta_{tr}G^T$) and Solvent Effects on Reaction Rates. Classification of Reaction Types**

| Case | $\delta_{tr}G^R$ | $\delta_{tr}G^T$ | Effect on rate[a] | Reaction type |
|------|------|------|------|------|
| 1 | − | − | +, 0 or − | Balanced |
| 2 | + | − | + | Positively reinforced |
| 3 | 0 | − | + | Positive transition-state control |
| 4 | − | 0 | − | Negative initial-state control |
| 5 | + | 0 | + | Positive initial-state control |
| 6 | 0 | 0 | 0 | Solvent independent |
| 7 | − | + | + | Positively reinforced |
| 8 | + | + | +, 0 or − | Balanced |
| 9 | 0 | + | − | Negative transition-state control |

[a] The plus sign refers to rate acceleration, the minus sign to rate retardation, and zero to no effect (Buncel and Wilson, 1979).

$$\Delta V = -RT\left(\frac{\partial(\ln K)}{\partial P}\right)_T \tag{1.28}$$

The analogous effect of pressure on the rate constant gives the volume of activation, $\Delta_{\ddagger}V$, through Equation 1.29. Measurement of reaction rates at high pressures, as Tobe and Burgess (1999, p. 10) point out, requires specialized apparatus; nevertheless, a great many volumes of activation are now available: see Isaacs (1984), van Eldik and Hubbard (1997), van Eldik and Meyerstein (2000), van Eldik et al. (1989), Blandamer and Burgess (1982), Laidler (1987), Tobe and Burgess (1999), and the large compilations of volumes of reaction and activation in the reviews by Drljaca et al. (1998 and references therein).

$$\Delta_{\ddagger}V = -RT\left(\frac{\partial(\ln k)}{\partial P}\right)_T \tag{1.29}$$

It is not likely that $\Delta_{\ddagger}V$ is large, though values outside $\pm 10\,\mathrm{ml\,mol^{-1}}$ have been obtained, notably for reactions consuming or generating ions in polar solvents. van Eldik and Meyerstein (2000) show that in favorable cases there is a linear correlation between $\Delta_{\ddagger}V$ and $\Delta V$. They present the example of substitutions on $Pd(H_2O)_4^{2+}$ in a variety of solvents, where $\Delta_{\ddagger}V \approx \Delta V - 2\,\mathrm{cm^3\,mol^{-1}}$. Where the activated complex resembles the products, this correlation is not unexpected, but it is by no means universal. Tobe and Burgess (1999) present *volume profiles*, which are schematic graphs of the volume changes along the reaction pathway, showing cases in which a degree of correlation exists (p. 536) and others in which it clearly does not (pp. 11, 301).

Volumes and entropies of activation for many classes of reactions show parallel trends, and can be interpreted in similar terms. In some cases the volume of activation is more reliable than the entropy of activation, because the latter is obtained by what may be a long extrapolation of the plot of $\ln k$ against $1/T$ to obtain the intercept.

The volume of activation for a reaction in an inert solvent can be a help in the assignment of a mechanism, because dissociative activation may be assumed to result in a positive volume of activation (in the region of $10–15\,cm^3\,mol^{-1}$ for each bond assumed to be stretching in the activation process), and associative activation the reverse. In a solvent that interacts strongly with solutes, however, these interactions must also be taken into account. Reactions in which ions are generated, such as Menschutkin reactions (e.g., 1.30), are characterized in solution by large, negative, and solvent-dependent entropies and volumes of activation: $\Delta_{\ddagger}V=-12$ to $-58\,cm^3\,mol^{-1}$ (Tobe and Burgess, 1999), because solvation of the nascent ions leads to reduced solvent freedom, reducing the entropy, and at the same time electrostriction of the solvent, reducing the volume.

$$Et_3N+Et-I\rightarrow[Et_3N^{\delta+}\text{---}Et\text{---}I^{\delta-}]^{\ddagger}\rightarrow Et_4N^++I^- \tag{1.30}$$

## PROBLEMS

**1.1** Data are tabulated for the equilibrium in aqueous solution between $N,N'$-bis-(hydroxymethyl)-uracil, **A**, and methanol, **B**, to form the diether, **C**, and water:

$$\mathbf{A}+2\mathbf{B}\rightleftharpoons\mathbf{C}+2H_2O$$

(Reagents were mixed in stoichiometric amounts.)

| $C_0$ | 0.10 | 0.25 | 0.5 | 1 |
|---|---|---|---|---|
| $f$ at 17°C | 0.0125 | 0.066 | 0.18 | 0.35 |
| $f$ at 30°C | 0.018 | 0.091 | 0.23 | 0.40 |

$C_0$=initial concentration of **A** in $mol\,l^{-1}$.

$f$=fraction converted at equilibrium.

**(a)** Calculate the mean equilibrium constant at each temperature (using the convention that the activity of water, approximately constant, is unity).

**(b)** Calculate $\Delta G^0$ at each temperature, and assuming they are constant, $\Delta H^0$ and $\Delta S^0$.

**(c)** With the same assumption, calculate the equilibrium constant and the fraction converted at 100°C, $C_0=1.0$. (Hint: Try successive approximation, use a calculator with a "Solve" program, or solve graphically.)

**1.2** Use the values of enthalpy of formation and entropy given here to calculate the equilibrium constant at 25°C for the esterification reaction in the vapor phase:

$$CH_3COOH+C_2H_5OH=CH_3COOC_2H_5+H_2O$$

Does it make any difference to the numerical value whether the constant is expressed in mole fraction, concentration, or pressure units?

(a) Use the vapor pressures of the pure substances given to calculate the equilibrium constant (in mole fraction terms) for this reaction in a solvent in which all four substances form ideal solutions (a practical impossibility).

| Substance (all as gas) | $\Delta_f H^0_{298} \, kJ^{-1} mol^{-1}$ | $S^0_{298} \, J^{-1}K^{-1}mol^{-1}$ | $p^0_{298} \, mm^{-1}Hg$ |
|---|---|---|---|
| Acetic acid | −434.3 | 282.7 | 15.4 |
| Ethanol | −235.37 | 282 | 57.2 |
| Ethyl acetate | −437.9 | 379.6 | 90.5 |
| Water | −241.8 | 188.83 | 23.8 |

**1.3** Pure $p$-xylene and water were equilibrated at 25°C. The absorbance $A_0$ of the aqueous layer measured in a 1 cm cell at $\lambda_{max} = 274 \, nm$ (due to $p$-xylene) was 0.884. A solution of $p$-xylene, mole fraction $x_1 = 0.686$, and $n$-dodecane, similarly treated, gave absorbance $A = 0.749$. Assuming that both Beer's and Henry's laws hold for $p$-xylene in water and that $n$-dodecane is insoluble in water, what was the activity coefficient of $p$-xylene in the solution with $n$-dodecane? (Neglect the small solubility of water in $p$-xylene.)

**1.4** The rate constants of the reaction between OH radical and $H_2$ in the gas phase at 25, 45, and 100°C were found to be $3.47 \times 10^3$, $1.01 \times 10^4$, and $1.05 \times 10^5 \, l \, mol^{-1} s^{-1}$, respectively. (a) What are the Arrhenius activation energy $E_a$ and the pre-exponential factor $PZ$ at 25°C? (b) Calculate the activation equilibrium constant $K^{\ddagger}$ and the activation parameters $\Delta_{\ddagger}H^0$, $\Delta_{\ddagger}S^0$, and $\Delta_{\ddagger}G^0$. (c) If the collision diameters of OH and $H_2$ are 310 and 250 pm, calculate the collision number $Z$ and obtain an estimate of $P$ at 25°C. (Caution: watch the units!)

# 2

# UNREACTIVE SOLVENTS

## 2.1 INTERMOLECULAR POTENTIALS

In this chapter we consider those aspects of the interaction of solvent and solute that are most clearly "physical" in nature, setting aside the more "chemical" aspects until the next chapter. The intermolecular potential characteristic of nonpolar substances, the London or dispersion interaction, arises from the mutual, time-dependent polarization of the molecules. For two molecules well separated *in vacuo* it is approximated by the London formula:

$$V = -\frac{\alpha_1' \alpha_2' I_h}{3r^6} \tag{2.1}$$

where the factors $\alpha_1'$, $\alpha_2'$ are the polarizability volumes of the molecules, $I_h$ the harmonic mean of their ionization energies, and $r$ is the separation of their centers. Equation 2.1 obviously is not accurate for a liquid, but it gives a sufficiently close estimate to enable Atkins (1998, p. 665) to conclude that the dispersion interaction is the dominant one in all liquids where hydrogen bonding is absent.

The remaining interactions in molecules that have permanent dipole moments, that is, the *dipole–dipole* and *dipole–induced-dipole* interactions, have the same dependence on intermolecular separation as the London potential, varying as $r^{-6}$, and are of lesser magnitude at ordinary temperatures (Atkins, 1998). These two interactions and the previously mentioned dispersion interaction are collectively known as *van der Waals interactions*. They are related to such measurable properties as surface tension and energy of vaporization, and to concepts such as the internal pressure, the *cohesive energy density* (energy of vaporization per unit volume, $\Delta_{vap} U/V$), and the *solubility parameter*, $\delta$, which is the square root of the cohesive energy density (Hildebrand and Scott, 1962).

*Solvent Effects in Chemistry*, Second Edition. Erwin Buncel and Robert A. Stairs.
© 2016 John Wiley & Sons, Inc. Published 2016 by John Wiley & Sons, Inc.

The *polarity* of the molecules is usually considered to be measured on a gross scale by the relative permittivity and on a molecular scale by the electrical dipole and higher moments. Molecules lacking a dipole moment (carbon dioxide, for example) may still exert short-range effects due to quadrupole, and so on, moments. Dipolar bonds that are well separated in a molecule may act almost independently on neighboring molecules; Hildebrand and Carter (1930) showed that the three isomeric dinitrobenzenes, in their binary solutions in benzene, exhibit nearly identical deviations from Raoult's law, though their dipole moments are different. The part of the electrical influence of a solvent on solute molecules that arises from the polarizability of the solvent molecules may be represented by the refractive index, $n$, or by functions of $n$ such as the *volume polarization*, $R$, given by:

$$R = \frac{n^2 - 1}{n^2 + 2} \tag{2.2}$$

([$R$], the molar polarization, is $R$ multiplied by the molar volume.) To allow for distortion polarization the refractive index should be that for far-infrared radiation, but this is not usually known, so the usual sodium D-line value is commonly used. The corresponding quantity including the effects of the permanent dipole moment is of the form $P = (\varepsilon_r - 1)/(\varepsilon_r + 2)$, where $\varepsilon_r$ is the relative permittivity, or dielectric constant. A possible measure of polarity, as distinct from polarizability, may then be defined by $Q = P - R$, that is, by Equation 2.3:

$$Q = \frac{\varepsilon_r - 1}{\varepsilon_r + 2} - \frac{n^2 - 1}{n^2 + 2} \tag{2.3}$$

Sometimes, as when hydrogen bonding is possible, a more detailed charge distribution is important. This, however, is approaching the region of specific chemical effects, so will be deferred to the next chapter. Considerations of polarity become paramount when a solute is an ionic substance. We will consider first the category of relatively nonpolar, nonelectrolyte solutions, then polarity, and finally, solutions of electrolytes in molecular solvents.

## 2.2 ACTIVITY AND EQUILIBRIUM IN NONELECTROLYTE SOLUTIONS

Let us consider a general reversible reaction in solution:

$$\mathbf{A}(\text{solv}) + \mathbf{B}(\text{solv}) \rightleftharpoons \mathbf{C}(\text{solv}) + \mathbf{D}(\text{solv}) \tag{2.4}$$

("$\mathbf{A}(\text{solv})$" represents a molecule of $\mathbf{A}$ surrounded by, but not necessarily strongly interacting with, molecules of the solvent.) One may rigorously write down the expression for the thermodynamic equilibrium constant:

$$K = \frac{a_C a_D}{a_A a_B} \tag{2.5}$$

where the *a*s are activities. If the solution is ideal for all components, a highly unlikely event, all the activities may be replaced by mole fractions *x*:

$$K_x = \frac{x_C x_D}{x_A x_B} \tag{2.6}$$

This generally untrue statement can be rehabilitated by multiplying the mole fractions by *rational activity coefficients, f*, to reconvert them to activities:

$$K = \frac{x_C f_C x_D f_D}{x_A f_A x_B f_B} = K_x \frac{f_C f_D}{f_A f_B} = K_x \Gamma_f \tag{2.7}$$

so that if we can calculate or measure the activity coefficients, we can then use the activity coefficient quotient $\Gamma_f$ to predict the effect of the solvent on the value of the *practical equilibrium constant* $K_x$.

There is a certain freedom of choice still open to us as to the choice of the conditions in which the *true* or *thermodynamic equilibrium constant* is defined. Commonly, in working with solutions in a single solvent, one uses a definition according to which all activity coefficients become unity in the limit of extreme dilution. This is very convenient in that the practical and thermodynamic equilibrium constants become in the limit identical, and may not differ too much at moderate dilution. It will not do here, though, for we wish to focus attention on the changes that result from a change of solvent, even if the solutions in both solvents are exceedingly dilute. We therefore must choose a standard state for each solute that is the same regardless of the solvent, and hence a single *K* that is a function of temperature alone. The limit that each practical constant, $K_x$, $K_p$, or $K_C$, approaches as the concentrations are decreased will still depend on the properties of the solvent. The simplest choice for this single *K* is that for the reaction in the gas phase at low pressure (so the Ideal Gas Law applies to all species), with no solvent present:

$$A(g) + B(g) \rightleftharpoons C(g) + D(g): \quad K = K_p$$

We now imagine a solvent in which all the solutes form ideal solutions, and applying Raoult's law (Eq. 1.5) to each *p* in $K_p$ (Eq. 1.4 rewritten for this reaction, with $\Gamma_\gamma = 1$), we obtain:

$$K_p = \frac{p_C^0 x_C p_D^0 x_D}{p_A^0 x_A p_B^0 x_B} = K_x(\text{ideal}) \frac{p_C^0 p_D^0}{p_A^0 p_B^0} \tag{2.8}$$

Since the $p^0$ s, the vapor pressures of the several pure substances, depend only on temperature, $K_x(\text{ideal})$ is a true constant, just as good as $K_p$. This is probably the most convenient candidate for the thermodynamic constant, *K*, in all cases of interest to us except those involving ionic species. Unless a comment is made to the contrary, it may be assumed that $K = K_x(\text{ideal})$ in the following pages.

Deviations of $K_x$ from $K$ can be of three sorts. Deviations at low and moderate concentrations from the zero-concentration limiting value are most important where ionic solutes are dissolved in nonionic solvents. Where all the solutes are molecular, even at extreme dilution, departures from Raoult's law on the part of each solute can arise in two distinct ways, with opposite effects. Deviations that are due to specific chemical or quasi-chemical attractive interactions between unlike molecules and that lead to enhanced mutual solubilities, lower partial vapor pressures, and activity coefficients less than unity are called *negative deviations*. Those that arise from mere differences between the molecules of the two kinds, such as differences of size or shape or of the intensity of intermolecular forces (reflected in differences in the solubility parameter, defined later), and that lead to diminished solubility, higher partial vapor pressures, and activity coefficients greater than unity are called *positive deviations* (see Fig. 1.2).

The effect of mere difference between molecules, such as different size, shape, or polarizability, as it affects intermolecular forces was considered by Hildebrand and Scott (1962). They were led to the concepts of *cohesive energy density* (which is defined as the molar energy of evaporation divided by the molar volume, and which has the dimensions of pressure) and its square root, the *solubility parameter*, defined by:

$$\delta = \left( \frac{\Delta U_{vap}}{V_{mol}} \right)^{1/2} \tag{2.9}$$

When two liquids are mixed, if the molecules distribute themselves randomly (thermal agitation being enough to overcome any tendency to specific pairing or segregation), the resulting solution is termed "regular." Assuming regular solution, and neglecting any volume changes, Hildebrand and Scott derive the relation:

$$RT \ln a_2 = RT \ln x_2 + V_2 \varphi_1^2 \left( \delta_2 - \delta_1 \right)^2 \tag{2.10}$$

in which $a_2$, $x_2$, $V_2$, and $\delta_2$ are the activity, the mole fraction, the molar volume, and the solubility parameter of component **2** in the mixture and $\varphi_1$ and $\delta_1$ are the volume fraction and the solubility parameter of component **1**. If the solution were ideal $a_2$ would be equal to $x_2$ so the last term is a measure of the departure of the solution from ideality. Note that in pure **2**, where $\varphi_1$ is 0, $a_2 = x_2 = 1$; in all very dilute solutions the solvent behaves as if the solution were ideal. In terms of the rational activity coefficient, $f_2$:

$$RT \ln f_2 = V_2 \varphi_1^2 \left( \delta_2 - \delta_1 \right)^2 \tag{2.11}$$

In a very dilute solution of **2** in **1**, on the other hand, $\varphi_1 \approx 1$, so $f_2$ is constant, and the solute obeys Henry's law. Very great approximations were used in the derivation of these equations, but they are surprisingly good as long as the components of the mixture do not carry too large dipole moments or take part in hydrogen bonding.

Now let us consider the reaction 2.4 in dilute solution in a solvent **S**:

We will use Equation 2.11 four times, putting **A**, **B**, and so on, in turn as component **2**, and **S** as component **1**, obtaining Equation 2.12, and we may write the practical equilibrium constant: $K_x = K/\Gamma_f$.

$$\ln \Gamma_f = \frac{f_C f_D}{f_A f_B} = \frac{\varphi_S^2}{RT} \left[ V_C \left( \delta_C - \delta_S \right)^2 + V_D \left( \delta_D - \delta_S \right)^2 - V_A \left( \delta_A - \delta_S \right)^2 - V_B \left( \delta_B - \delta_S \right)^2 \right]$$

$$= \frac{\varphi_S^2}{RT} \Delta \left[ V \left( \delta - \delta_S \right)^2 \right]$$

$$(2.12)$$

Since the thermodynamic constant $K$ is a good constant, depending only on the temperature, anything that decreases $\Gamma_f$ will increase $K_x$, and will increase the relative amounts of **C** and **D** present at equilibrium. This will be the case if the $(\delta - \delta_S)^2$ terms are kept small for **C** and **D**, and large for **A** and **B**. That is to say, if we want a good yield of **C** or **D**, we should carry out the reaction in a solvent as much like **C** or **D**, or both, as possible, and unlike **A** and **B**, in the sense of the saying "Like dissolves like."

For example, consider the reaction:

$$CO + Br_2 \rightleftharpoons COBr_2$$

(a somewhat artificial example, for the equilibrium lies too far to the right for convenient measurement in solution at room temperature, but it illustrates the method in a case where the result is not obvious without calculation). The relevant quantities, assuming the partial molal volumes of the solutes are equal to their molal volumes in the pure liquids, are:

|  | CO | $Br_2$ | $COBr_2$ |
|---|---|---|---|
| $\delta/MPa^{1/2}$ | 13.5 | 23.7 | 17.4 |
| $V/ml\,mol^{-1}$ | 35 | 55 | 92 |

Equation 2.12 with these values leads to:

$$\ln \left( \frac{K_x}{K} \right) = 3.80 - 0.1414\ \delta_S - 0.0008\ \delta_S^2 \qquad (2.13)$$

It would have been hard to guess whether the increase in $\delta$ from CO to the product, or the decrease from $Br_2$ to the product, would dominate the effect. As a rough estimate, where the volume change of the reaction is small (here $\Delta V = 92 - 35 - 55 = 2\,ml$) the slope of the plot of $\ln(K_x/K)$ versus $\delta_S$ is approximately equal to $2\Delta(V\delta)/RT$, here about $-0.14$. The negative slope means that the formation of the product is favored in a medium of low solubility parameter.

## 2.3   KINETIC SOLVENT EFFECTS

The approach discussed earlier may be applied to the effect of solvent on the rate of a reaction, through the thermodynamic interpretation of the Transition State Theory (see Chapter 1). Representing a reaction as:

$$A + B \rightleftharpoons \ddagger \rightarrow products,$$

we apply Equation 2.11 to the activated complex $\ddagger$ and to the reactants, as earlier, and obtain:

$$
\begin{aligned}
\ln \frac{k}{k_0} &= \ln \frac{1}{\Gamma_f} = \ln \frac{f_A f_B}{f_\ddagger} \\
&= \frac{\varphi_S^2}{RT} \left[ V_A \left( \delta_A - \delta_S \right)^2 + V_B \left( \delta_B - \delta_S \right)^2 - V_\ddagger \left( \delta_\ddagger - \delta_S \right)^2 \right] \qquad (2.14) \\
&= -\frac{\varphi_S^2}{RT} \Delta_\ddagger \left[ V \left( \delta - \delta_S \right)^2 \right] \\
&= -\frac{\varphi_S^2}{RT} \left[ \Delta_\ddagger \left( V \delta^2 \right) - 2 \Delta_\ddagger (V \delta) \delta_S + \Delta_\ddagger V \delta_S^2 \right]
\end{aligned}
$$

The molar volumes of the reactants, $V_A$, $V_B$, are presumably known. The $\delta$s are known for many substances or can be calculated from known quantities (Hildebrand and Scott, 1962; Marcus, 1985). The molar volume of the activated complex is not usually known; $\Delta_\ddagger V$ may be obtained through the effect of pressure on the reaction rate (see earlier, Section 1.10). Finally, $\delta_\ddagger$ may be obtained from the best fit of data for the rate constant of the reaction in a variety of solvents to Equation 2.14.

An example occurred in a study of the rate of oxidation of toluene by chromyl chloride (Étard, 1881) in assorted solvents (Stairs, 1962; Cook and Meyer, 1995):

$$C_6H_5CH_3 + CrO_2Cl_2 \rightarrow C_6H_5CH_2OCrOHCl_2 \quad k_1 \, (\text{slow})$$

$$C_6H_5CH_2OCrOHCl_2 + CrO_2Cl_2 \rightarrow C_6H_5CH(OCrOHCl_2)_2 \quad k_2 (\text{somewhat faster})$$

The structures of the intermediate and of the final product, a brown precipitate insoluble in all solvents that do not destroy it, are conjectural. The latter appears to contain Cr(IV).[1] On hydrolysis, benzaldehyde, some benzyl chloride, HCl, and a mixture of Cr(III) and Cr(VI) species are formed.

In Figure 2.1a the logarithm of the observed rate constant $k_1$ is plotted against the solubility parameters of the six solvents. The curve was drawn with the assumed value $-10 \, \text{cm}^3 \, \text{mol}^{-1}$ for $\Delta_\ddagger V$ and fitted to all the points but one (more about that one later). The moderate increase in rate with increase in $\delta_S$ was taken as evidence that the activated complex had a higher solubility parameter than the reactants, but not

---

[1]This was supported by magnetic susceptibility. An attempt to obtain an electron spin resonance spectrum failed, owing to the extreme broadening due to a two-electron system (Stairs, 1990, unpublished result).

(a)

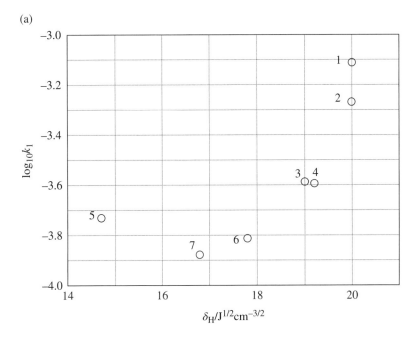

(b)

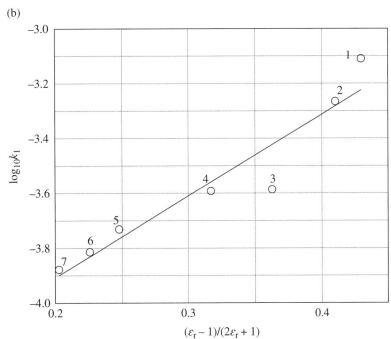

**FIGURE 2.1** (a) Kinetic solvent effect in Etard's reaction. $\mathrm{Log}_{10}(k_1)$ versus Hildebrand's solubility parameter. Solvents are (from Stairs, 1962): (1) 1,2-dichloroethane, (2) 1,1,2,2-tetrachloroethane, (3) trichloromethane, (4) pentachloroethane, (5) 1,1,2-trichloro-f-ethane (Freon 113), (6) tetrachloromethane, and (from Cook and Meyer, 1995) (7) cyclohexane. Redrawn after Stairs (1962) by permission of the National Research Council of Canada. (b) Logarithms of rate constants as in (a) versus Kirkwood's dielectric function. Numbers refer to solvents as in (a). The slope of the fitted line is $2.9\pm0.2$ (dimensionless).

much higher, and that it was probably not ionic in nature. This was helpful in an attempt to assign a mechanism to the reaction.

## 2.4 SOLVENT POLARITY

The preceding treatment was based on the assumption that none of the molecules involved is so polar as to exert strong orienting forces or specific attractive forces on neighboring molecules. It breaks down in cases such as propanone (acetone), which has a rather low solubility parameter based on its energy of vaporization, but which must be assigned a high and variable one to account for its ability to dissolve liquids such as water, and even inorganic salts. It dissolves these substances through specific interactions: hydrogen bonding with water molecules, and ion–dipole solvation of the cations of salts. A less extreme case is apparent in the kinetic data discussed in the last section. Point 5 in Figure 2.1a is for the solvent 1,1,2-trichloro-1,2,2-trifluoroethane (Freon 113), which has a low solubility parameter ($\delta = 14.7\,\text{MPa}^{1/2}$) owing to the weak van der Waals (London) forces typical of highly fluorinated molecules, but which is, nevertheless, somewhat polar ($\mu = 0.4$ Debye, approximately[2]). It has about the same effect on the rate as tetrachloromethane, which, though nonpolar, has higher London forces, leading to $\delta = 17.9\,\text{MPa}^{1/2}$.

It is possible to extend the solubility-parameter method to include the effects of moderate polarity by assuming the cohesive energy density ($\Delta_{vap} U/V$) to be made up of two parts, $\delta^2 + \omega^2$, where $\delta^2$ is a measure of the London interaction and $\omega^2$ (which is proportional to $\mu^4$) of the polar interaction. With this complication it becomes less convenient, however, and less satisfactory.[3]

## 2.5 ELECTROSTATIC FORCES

A very different approach from the foregoing was made by Kirkwood (1934; Amis and Hinton, 1973, p. 241; Onsager, 1936). Kirkwood looked at the molecules of the solute as spheres, each bearing at its centre an electric dipole moment $\mu$, in a continuous medium of relative permittivity $\varepsilon_r$. (The relative permittivity, or dielectric constant, is the ratio of the permittivity $\varepsilon$ of the medium to the permittivity of free space, $\varepsilon_0$.) The difference in free energy of a mole of such spherical dipoles in this medium from what it would be if the relative permittivity were unity is given by:

$$G_{\varepsilon_r} - G_1 = RT \ln f$$

$$= -\frac{L\mu^2}{4\pi\varepsilon_0 r^3} \frac{(\varepsilon_r - 1)}{(2\varepsilon_r + 1)} \tag{2.15}$$

[2]The Debye unit of dipole moment is equal to $10^{-18}$ esu cm, or $3.336 \times 10^{-30}$ C m. in SI.
[3]The reader may pursue the matter in treatments by Hildebrand and Scott (1950, chapter IX) and by Burrell (1955), and in the review by Barton (1975).

Here $f$ is the activity coefficient, $L$ is Avogadro's number, and $r$ is the radius of the sphere. Applying Equation 2.15 to each of the species in the reaction: $A + B \rightleftharpoons C + D$, we obtain:

$$RT \ln \Gamma_f = -\frac{N(\varepsilon_r - 1)}{4\pi\varepsilon_0 (2\varepsilon_r + 1)} \left( \frac{\mu_C^2}{r_C^3} + \frac{\mu_D^2}{r_D^3} - \frac{\mu_A^2}{r_A^3} - \frac{\mu_B^2}{r_B^3} \right) \tag{2.16}$$

Again recalling that $K_x = K/\Gamma_f$, to obtain a good yield of $C$ or $D$ we want $\Gamma_f$ to be small. This turns out to lead to the same rule of thumb as in the nonpolar treatment, that the best solvent is the one that most resembles the products. *If the desired products are more polar than the reactants, the reaction is favored by a polar solvent, but if the products are less polar, a relatively nonpolar solvent is preferred.*

Equation 2.16 may be applied to the reaction considered earlier, between bromine and carbon monoxide. Bromine is nonpolar. Carbon monoxide has a small dipole moment, of magnitude 0.112 Debye, and the product, carbonyl bromide, has a somewhat larger dipole moment, about 1.2 Debye. The mean radii of the three molecules approximated as spheres are as given: ($Br_2$) 0.23 nm, (CO) 0.20 nm, ($COBr_2$) 0.296 nm. If these values are used in Equation 2.16, the result is:

$$\log_{10}\left(\frac{K_x}{K}\right) = 0.566 \frac{\varepsilon_r - 1}{2\varepsilon_r + 1} \tag{2.17}$$

As would be expected, the formation of the more polar product is favored in solvents of higher relative permittivity (contrary to the prediction based on the solubility parameters).

Solvents of low relative permittivity tend to also have low solubility parameters. Figure 2.2 shows a plot of the relative permittivity function against the solubility parameter for 26 solvents of varied character, including the six solvents used in the study of Étard's reaction, given earlier (marked in the figure by crosses). Since the correlation between these two properties is clearly weak, it should be possible in some cases to tell whether the polarity or the nonpolar (London) part of the intermolecular forces has the more important part in determining the total effect of the solvent on the yield of product. For the reaction of carbon monoxide with bromine, the two theories predict opposite effects, so it is unfortunate that this reaction is a difficult one to study.

As in the previous treatment, this method can also be applied to reaction rates, through the thermodynamic interpretation of the Transition State Theory. Figure 2.1b shows the result of applying Equation 2.16 to the same kinetic data for Étard's reaction, discussed earlier in terms of solubility parameters. Here the rate constants are plotted (as common logarithms) against $(\varepsilon_r - 1)/(2\varepsilon_r + 1)$. Comparing this figure to Figure 2.1a, it is immediately apparent that the point for the Freon, which is well off the trend in (a), is close to the line in (b), suggesting that the electrostatic part of the effect is more important, though the parallel trends in the solubility parameters and the dielectric functions of the rest of the solvents make it difficult to confirm this

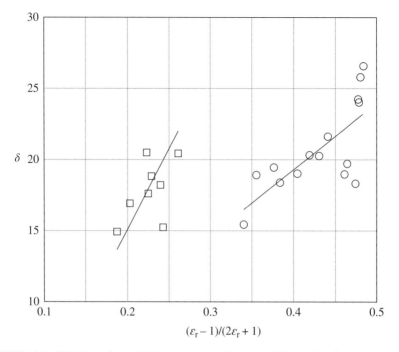

**FIGURE 2.2** Hildebrand's solubility parameter $\delta$ versus Kirkwood's dielectric function $(\varepsilon_r - 1)/(2\varepsilon_r + 1)$ for a selection of solvents of low (squares) and high (circles) dipole moment.

conclusion. (It was difficult to find solvents with which chromyl chloride did not react.) The dipole moment of the activated complex may be estimated from the slope of the line in Figure 2.1b as about 2.0 Debye, which is a value typical of rather polar molecules.

As another example, Reichardt (1988, p. 154) cites data from Huisgen and coworkers (Swieton *et al.*, 1983) for the cycloaddition of diphenylketene to *n*-butyl vinyl ether. Figure 2.3 shows a plot of the natural logarithm of the rate constant (relative to the slowest) versus the Kirkwood function. The slope corresponds to a dipole moment for the transition state of about 10 Debye ($34 \times 10^{-30}$ C m), larger than that of the product, indicating a considerable degree of charge separation.

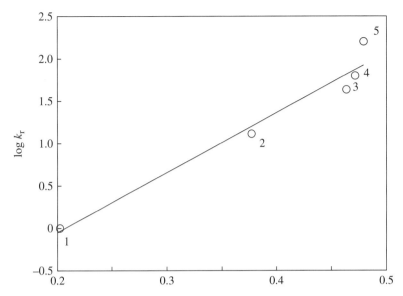

**FIGURE 2.3** Cycloaddition of diphenylketene to butyl vinyl ether. Common logarithm of the relative rate constant versus Kirkwood's dielectric function. Data from Reichardt and Weldon (2011, p. 181). Solvents were 1, cyclohexane; 2, chlorobenzene; 3, acetone; 4, benzonitrile; 5, acetonitrile.

## 2.6  ELECTROLYTES IN SOLUTION

When a typical salt is dissolved in a liquid, such as water, in which it is a strong electrolyte, the ions of the salt interact both with the solvent molecules and with each other. The latter interaction persists on dilution to very low concentrations, for the Coulomb force between like or oppositely charged ions extends to long distances. It depends on concentration in a way that can be calculated, at least in dilute solutions, by the Debye-Hückel (1923) Theory. This theory has been fully treated elsewhere (e.g., Atkins, 1998, pp. 248–253; Atkins and de Paula, 2010, pp. 196, 199; Barrow, 1988; Skoog *et al.*, 1989), so let us note its main features here.

The theory is commonly quoted at several levels of approximation. The simplest form, which is valid only at the lowest concentrations (below 0.001 mol l$^{-1}$ in water, and lower in most other solvents), is the *limiting law* (for a solution of a single salt):

$$\log_{10} \gamma_\pm = A z_+ z_- c^{1/2} \tag{2.18}$$

The constant $A$ depends on the relative permittivity of the solvent and on the temperature, in the form: $1.8246 \times 10^{-6} (\varepsilon_r T)^{3/2}$. For water at 25°C its value is $0.51151^{1/2}$ mol$^{-1/2}$. The *mean ionic activity coefficient*, $\gamma_\pm$, is defined so that the activity $a = \gamma_\pm c$ becomes equal to the molar concentration $c$ in the low concentration limit. The subscript $\pm$ is added to the symbol because it is not possible rigorously to define the activities of the separate ions, nor their activity coefficients. By assuming that the activity coefficients of the two ions of a symmetrical electrolyte are equal, however, it is possible to write a single-ion version:

$$\log_{10} \gamma = -Az^2 c^{1/2} \tag{2.19}$$

The negative sign appears because one of the $z$s in Equation 2.18 is negative.

For a solution containing more than two kinds of ions it is necessary to define the *ionic strength, I,* by the relation:

$$I = \frac{1}{2}\Sigma_i z_i^2 c_i \tag{2.20}$$

With this definition, the limiting law for a single ion becomes:

$$\log_{10} \gamma = -Az^2 I^{1/2} \tag{2.21}$$

To extend the application of the theory to more useful concentrations, the fact that two oppositely charged ions can approach each other only until their centers are separated by the sum of their radii, $a$, is used to correct Equation 2.21 to read:

$$\log_{10} \gamma = -\frac{Az^2 I^{1/2}}{1 + BaI^{1/2}} \tag{2.22}$$

The value of the new constant $B$ in water at 25°C is $3.291 \times 10^9 \, \text{m}^{-1} \, \text{mol}^{-1/2} \text{l}^{1/2}$. For many electrolytes the value of $a$ is around 300–400 pm, so the product $Ba$ is close to unity for aqueous solutions at room temperature. Taking advantage of this coincidence, and adding an empirical linear term in $I$ (which may be justified by some qualitative reasoning), Davies (1962) has formulated a useful approximate expression, Equation 2.23, which he finds applicable to a great many ionic species in water at room temperature.

$$\log_{10} \gamma_z = -0.5z^2 \left( \frac{I^{1/2}}{1 + I^{1/2}} - 0.3I \right) \tag{2.23}$$

Figure 2.4 illustrates the form of the limiting law (Eq. 2.19), the "extended" Equation (2.22) and Davies's approximation (2.23), all for a univalent ion in aqueous solution at 25°C.

At low and moderate concentrations, the theory predicts that the logarithm of the activity coefficient will be negative, and inversely dependent on the product of the temperature and the relative permittivity of the solvent. At the lowest concentrations the dependence is on $(\varepsilon_r T)^{-3/2}$, but at moderate concentrations the exponent approaches $-1$. The effect of interionic forces at moderate ionic strengths is to favor

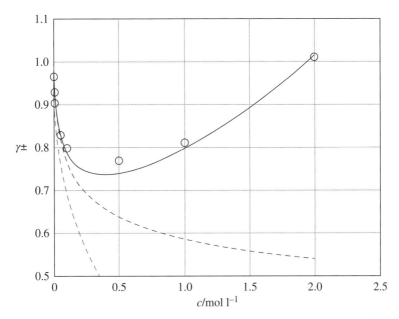

**FIGURE 2.4** The Debye–Hückel limiting law for a 1:1 electrolyte (Eq. 2.21, lower dashed curve), the corrected law (Eq. 2.22, upper dashed curve), Davies's approximation (Eq. 2.23, black curve) and experimental data for HCl, all in water at 25°C. The activity coefficient is plotted against ionic strength, $\mu = \Sigma z_i^2 c_i/2$. Data from Atkins and de Paula (2010, p. 927).

the formation of ionic products, for $\gamma_\pm$ in this region is always less than unity, but this effect is diminished as the relative permittivity increases. This diminution appears to contradict our expectation that a more polar solvent would favor the formation of ionic products, for ionization may be viewed as polarity carried to the extreme. It is, however, a small correction on the larger interaction of the ions with the solvent, discussed later.

As the concentration is increased further, deviations from the simpler forms of the Debye–Hückel Theory accumulate. The concentration beyond which Equation 2.22 fails depends strongly on the quantity $\varepsilon_r T$. In water at room temperature, where $\varepsilon_r T$ is about 24,000, the useful limit is in the range $0.01 - 0.1\,\text{mol}\,\text{l}^{-1}$ for uni-univalent salts, lower for higher valences. In liquid ammonia at its boiling point, $-33°C$, $\varepsilon_r T \approx 4800$, it is less than $0.001\,\text{mol}\,\text{l}^{-1}$. Furthermore, in solvents of low relative permittivity the ions tend to associate in pairs or higher aggregates (Section 2.9), so that in many solvents no strong electrolytes exist. Nevertheless, the theory is accurate over a limited range of concentrations, and (in the form of Davies's approximation) a useful guide over a wider one. It provides a mathematical form to aid extrapolation of certain data to very low concentrations, where the effects of interionic forces may become negligible, and the ion–solvent interactions may be separately examined.

## 2.7  SOLVATION

When the ions are so diluted that their mutual influences may be neglected, there still remains their interaction with the solvent, which is termed "solvation" (not "salvation", as a Calgary newspaper once headlined a report of a conference on this topic).[4] Solvent molecules may interact so strongly with a dissolved ion that they become firmly bound by ion–dipole forces, or by covalent bonds, and for certain purposes may be counted as part of the ion. Attempts have been made by various means to determine the number of molecules so bound, the *solvation number*. Different methods give different results, which should not be surprising, for the meaning of "firmly bound" depends on how hard one tries to dislodge them. Solvation numbers of ions in various solvents as found by different methods are discussed by Hinton and Amis (1971). Methods that have been used include measurement of apparent hydrodynamic radii of ions in diffusion and electrical conductance, of the amounts of one solvent (usually water) carried into another immiscible solvent with the solute in an extraction process, and of the amounts of solvent incorporated in crystals as "solvent of crystallization," though crystals may include solvent molecules not bound to any ion, but occupying sites elsewhere in the lattice. In $KAuBr_4.2H_2O$, for example, the water molecules are not attached to any ion, but occupy otherwise vacant spaces in the lattice (Cox and Webster, 1936). Various methods tend to suggest that sodium chloride in aqueous solution, for instance, carries between four and eight water molecules about the $Na^+$ ion, and fewer about the $Cl^-$ ion. Because the bound molecules are fully polarized, in the sense of being aligned with the ionic field, the inner sphere they occupy has been called the *sphere of dielectric saturation.*

The region about a solute particle (molecule or ion) within which the structure of the solvent is altered from what it is in the pure solvent, or in solution remote from a solute particle, is called the *cybotactic region* (see Fig. 2.5). About an ion one may distinguish the inner or coordination sphere, within one solvent molecular diameter, where solvent molecules are more or less firmly bound and strongly oriented, depending on the charge-to-size ratio of the ion and the polarity of the solvent molecules. Here the solvent structure is essentially destroyed, and the ion with its bound solvent acts as a larger ion. In the next region the solvent molecules, if dipolar, are oriented by the ionic field in diminishing degree as their distance from the ion increases. The conflict between this orientation and the tendency of the solvent to assume its normal structure may result in a more chaotic structure in this region, reflected, for instance, in diminished viscosity. Both the entropy of solution and the partial molar volume of the solute are decreased by the binding in the inner sphere, and increased by the disorder in the outer sphere. Which effect predominates depends on the size and charge of the ion.

The presence of a nonpolar solute in a polar solvent creates a rather different situation, described, in discussing aqueous solutions, as the *hydrophobic effect*, illustrated in Figure 2.5b. Here the water molecules, unable to form hydrogen bonds with the solute, do so with each other, but in a way that creates a cavity within which the

---

[4]On the other hand, E. S. Ames and J. F. Hinton (1973) dedicated their book, "*Solvent Effects on Chemical Phenomena*", vol. 1, New York and London, Academic Press, "To Dr. E. A. Moelwyn-Hughes, retired general of the Solvation Army".

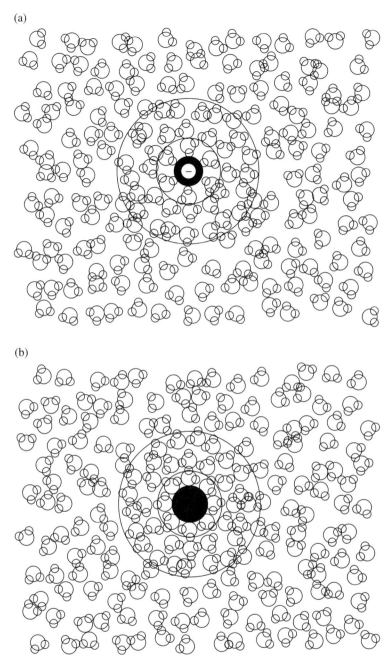

**FIGURE 2.5** (a) Solvation in flatland. A solvated anion (black circle) and its cybotactic region. The solvent molecules within the innermost circle are virtually fixed in orientation toward the ion. Those within the next circle are less strongly oriented, but more closely packed, while those beyond are undisturbed. (b) Hydrophobic solvation. The innermost solvent molecules form a cage around the nonpolar solute, hydrogen-bonded to each other. As in the ionic case, these are surrounded by a region of disturbed structure, beyond which the solvent is normal.

solute is contained. A quasi-crystalline inner sphere is formed, different in structure from the bulk water structure. Outside the inner sphere is again a chaotic region, before normal solvent structure is resumed at greater distances. Okazaki *et al.* (1979), in a Monte Carlo simulation (see Chapter 5) of methane in water, were able to distinguish these three regions. Both the enthalpy and the entropy are decreased, so the free energy change may have either sign. Abraham (1982) illustrated this by compiling data on the free energy, enthalpy, and entropy of solvation of 27 nonpolar gaseous substances in a number of solvents.

When two molecules thus solvated approach one another closely, the two cavities may merge, forming a single larger cavity, and liberating some of the solvent. The energy may increase or decrease, depending on details of hydrogen bonding; the entropy will increase. The entropy increase usually dominates, so dimerization is favored, that is, the equilibrium constant for the reaction depicted in Figure 2.6 is greater than unity.

If the solute contains ions such as carboxylate, with a charged head and a nonpolar tail, both ionic coordination and the hydrophobic effect can be simultaneously present. Soaps, such as sodium stearate $CH_3(CH_2)_{16}COO^-Na^+$, and detergents, such as sodium dodecylsulfate $CH_3(CH_2)_{11}OSO_3^-Na^+$, have long nonpolar tails. In aqueous solution above a certain concentration, the *critical micelle concentration*, they form *micelles*, more or less spherical globules in which the tails are together in the interior, and the polar or ionic heads on the surface. Figure 2.7 illustrates a spherical micelle.

The hydrophobic effect can lead in the extreme case to the formation of crystalline hydrates, such as the well-known chlorine hydrate $Cl_2.8H_2O$, and $CH_4.nH_2O$, found at great depth in ocean sediments, and recently of interest (Kleinberg and Brewer,

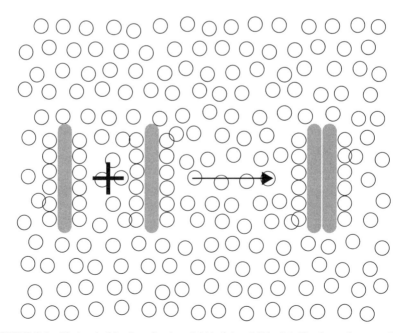

**FIGURE 2.6** Hydrophobic dimerization: $2M(solv) \rightarrow M2(solv)$. Twelve solvent molecules are shown as being liberated.

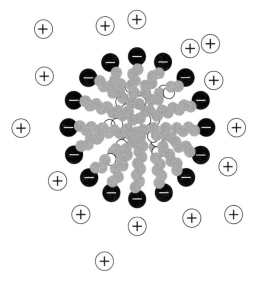

**FIGURE 2.7**   Schematic cross section of a micelle formed by an anionic detergent. The counterions shown would realistically be part of the ionic atmosphere about the micelle.

2001) as a large potential source of energy, and of concern lest climate change lead to large-scale melting and release of methane, which is a potent greenhouse gas. These are examples of the general class of substances called *clathrates*.

## 2.8   SINGLE ION SOLVATION

It is not possible by thermodynamic arguments alone to decide how the volumes of the ions and their solvation spheres are to be partitioned between the cation and anion of a dissolved salt. The conventional solution of this problem, as far as aqueous solutions are concerned, has been to make an arbitrary assignment of zero to the volume of the hydrogen cation. This is obviously a fiction, because the proton in water is known to be associated with at least one water molecule to form the "hydronium" or "hydroxonium" ion, $H_3O^+$, or more, as $H_9O_4^+$, (**1**) for example.

1

Nevertheless, the additivity of ionic properties in dilute solution ensures that all the relevant properties of the solute are correctly represented. Trémillon (1974) and Marcus (1985, pp. 96–105) present discussions of the various ways that partitioning of this and other ionic properties may be made, using arguments outside thermodynamics. The large size and similar structure of tetraphenylarsonium (**2**) and tetraphenylborate (**3**) ions have led to suggestions that they should have the same size in most solvents, and be nearly equally solvated.

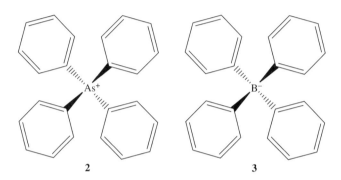

**2**                                                **3**

The conclusion Marcus comes to is that the best value for the partial molar volume of the hydrogen cation in extremely dilute aqueous solution is $-6.4 \, \mathrm{cm^3 \, mol^{-1}}$. The reason for this negative value is that by tightening the structure of water in its vicinity, through enhanced hydrogen bonding, and in more subtle ways (encompassed in the term *electrostriction*), the proton causes a local shrinkage. Similar effects are seen with many ions, especially those with small crystal radii or large charges. Other properties, such as enthalpies, Gibbs energies, and entropies may be partitioned by similar arguments (Cox, 1973; Marcus, 1985, pp. 105–113).

Outside the primary sphere is one in which the molecules are somewhat polarized, but not intensely so. If we take the radius of the ion $r$ to be the radius of the primary sphere, we may follow Born (1920) and Hunt (1963) and calculate the free energy of transfer of an ion of radius $r$ from one medium to another (let us assume that one is water and the other a solvent **S**) according to Equation 2.24, or for a mole of a 1:1 electrolyte, Equation 2.25:

$$\Delta G = -\frac{z^2 e^2}{8\pi\varepsilon_0}\left(\frac{1}{\varepsilon_r(w)} - \frac{1}{\varepsilon_r(S)}\right)\frac{1}{r} \tag{2.24}$$

$$\Delta G_{tr} = -\frac{Ne^2}{8\pi\varepsilon_0}\left(\frac{1}{\varepsilon_r(w)} - \frac{1}{\varepsilon_r(S)}\right)\left(\frac{1}{r_+} + \frac{1}{r_-}\right) = RT\Delta\ln\gamma_\pm^2 \tag{2.25}$$

Numerical calculation of these three effects, that is, the *primary and secondary solvation* and the interionic interaction, unfortunately increases in difficulty in the order of increasing importance. As Hunt shows, following Latimer *et al.* (1939), one may allow for the difficult-to-calculate primary solvation for a series of similar ions in a single solvent by assigning an "effective radius" and applying the Born equation. To approach this difficult situation experimentally, consider a reaction among a number of ionic or polar solutes in a medium consisting of a relatively nonpolar solvent containing some water, in which water molecules preferentially solvate the polar and ionic solutes. The primary solvation is by water and may be assumed to be constant. The remaining effects, attributed to the secondary solvation, may then be seen. Robinson and Stokes (1959) present a graph (redrawn as Fig. 2.8) in which the dissociation constants of a number of carboxylic acids in a number of aqueous mixed solvents are compared with those for the same acids in water. All the solvent mixtures contained enough water that the reaction in each case could be assumed to be:

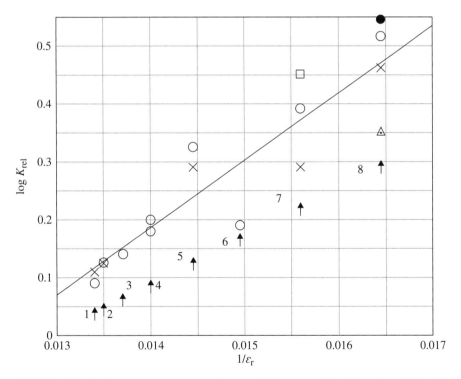

**FIGURE 2.8**    Acids in organic/aqueous solvents: (1) 2-propanol 5%, (2) methanol 10%, (3) ethanol 10%, (4) 2-propanol 10%, (5) methanol 20%, (6) ethanol 20%, (7) glycerol 50%, (8)1,4-dioxane 15%. Common logarithm of $K_a(S)/K_a(W)$ versus the reciprocal of the relative permittivity of the mixed solvent. Acids are indicated by plotting symbols: formic, triangle; acetic, cross; propanoic, circle; butanoic, square; water, filled circle. From Robinson and Stokes (1959) with permission of Dover Publications.

$$HA + (x + y + 1)H_2O \rightleftharpoons A^-.xH_2O + H_3O^+.yH_2O.$$

The dependence of $\log_{10}(K_S/K_W)$ on the reciprocal of the relative permittivity was clear, though a few points fall off the line. By neglecting any effect on the activities of the neutral species, we use Equation 2.25 to calculate from the slope (about $133 \pm 7$) of the line in the figure, $200 \pm 10$ pm, for the harmonic mean of the radii of hydrated $H^+$ and carboxylate. This seems somewhat small, but not impossible. The aberrant points may reflect failure of either of the assumptions: strongly preferential solvation by water or absence of medium effects on the unionized acid.

## 2.9   IONIC ASSOCIATION

The Debye–Hückel Theory has been very successful in explaining the behavior of strong electrolytes in solvents of relatively high relative permittivity, at low concentrations. The deviations noted earlier at higher concentrations are attributed to various short-range interactions between ions. In very concentrated solutions, the solution begins to be crowded, and begins to resemble a molten salt, with the beginnings of short-range order. This situation is not treatable by the methods contemplated here. At more moderate concentrations, oppositely charged ions may be significantly associated. In solvents with relative permittivities much less than that of water, hardly any electrolytes are strong (in the sense of being fully dissociated into separated, solvated ions). Thus in most solvents the study of electrolytes is mainly concerned with weak electrolytes. Quite aside from weak acids and bases, and the few salts, such as $HgCl_2$, that have genuine molecules, this nearly universal weakness in solvents of low relative permittivity needs explanation.

The problem has been approached by a number of workers, notable Bjerrum (1926), Gronwall et al. (1928), and Fuoss and Krauss (1933). These are discussed by Davies (1962). All agree that the relative permittivity of the solvent and the size and charge of the ions play large roles: *the lower the relative permittivity of the solvent, the smaller the ions (including tightly bound solvent molecules), and the larger the charges, the more a salt is associated into ion pairs*. There can be additional effects due to mutual polarization of the ions at short range, or even covalent bond formation. Bjerrum treats the solvent as a continuous dielectric, and derives a critical distance, $q = z_i z_j\, e^2/2\varepsilon_r kT$, at the minimum of the curve of probability for finding an ion of opposite charge as a function of distance from the central ion, seen in Figure 2.9. If the distance separating two ions ($r_{i,j}$) is less than $q$ the pair is "bound"; if greater it is free. If the distance of closest approach, $a$, exceeds $q$, pairing does not occur.

Fuoss and Krauss treat an ion pair as existing if the ions, including any solvent molecules firmly bound to the ions, are in contact, and not existing if one or more additional solvent molecules intervene. The probability of the ions being separated by a fraction of a solvent molecular diameter is low. Their resulting calculated dissociation constants are given by Equation 2.26.

$$K = \frac{\alpha^2 \gamma_\pm^2 c}{1 - \alpha} = \frac{3000}{4\pi N a^3 \exp b}; \quad b = \frac{|z_1 z_2| e^2}{\varepsilon_r kTa} \tag{2.26}$$

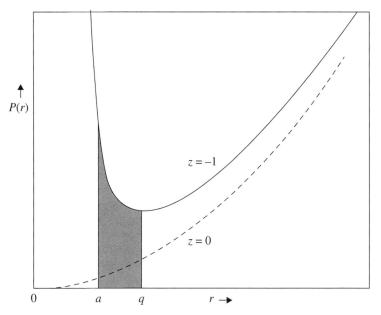

**FIGURE 2.9** Bjerrum probability distribution. Probability that a species of given charge (0 or −1) will be found at a distance $r$ from a central ion of charge +1. A negative ion inside the distance $q$ is considered paired with the central cation. If the distance of closest approach is $a$, the probability that a pair exists is proportional to the shaded area under the upper curve, $a < r < q$. If $a > q$, pairing does not occur. After Bjerrum (1926).

They are not so very different from Bjerrum's, which is a comfort, for explicitly accounting for solvent molecules is often difficult mathematically, and we should like to avoid it when we may. The difference is hard to test experimentally, but some of Fuoss and Krauss's data (1933) on the salt tetra-isoamylammonium nitrate suggest that their treatment is slightly better than Bjerrum's.

Winstein *et al.* (1954) distinguish contact ion pairs, solvent-separated ion pairs, and widely separated ion pairs, that is, free ions. The distinction between contact and solvent-separated ion pairs was demonstrated experimentally by Smithson and Williams (1958), Hogen-Esch and Smid (1966), and Buncel and Menon (1979; Buncel *et al.*, 1979). Figure 2.10a shows the visible absorption spectra of (triphenyl-methyl)lithium (**4**) in diethyl ether, in tetrahydrofuran, and in

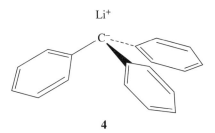

**4**

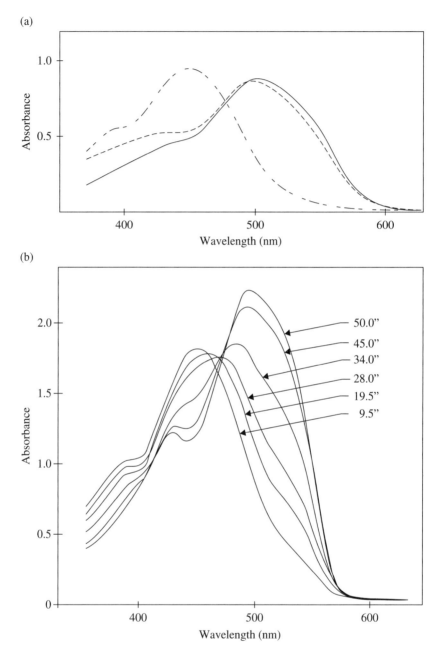

**FIGURE 2.10**   Visible absorption spectra of (triphenylmethyl)lithium: (a) at room temperature in diethyl ether (dash-dot), THF (solid) and 1,2-dimethoxyethane (dashes); (b) temperature dependence of the spectrum of the same substance in diethyl ether. Source: Buncel and Menon (1979). Reproduced with permission of the American Chemical Society.

1,2-dimethoxyethane at room temperature. The large shift of the absorption maximum was interpreted as indicating that the solute is present predominantly as contact ion pairs in diethyl ether, but solvent-separated in THF and in dimethoxyethane. Changes were observed in the absorption spectrum in diethyl ether as the temperature was lowered from $-9.5$ to $-50°C$ (see Fig. 2.10b).

The absorption maximum near 450 nm appears to be due to the contact ion pair and that near 500 nm to the solvent-separated pair. As the temperature is lowered, solvent-separated ion pairs are favored. Buncel and Menon calculate for $K$, the ratio of the concentrations of solvent-separated to contact pairs, the value 0.1 at 25°C, and estimate $\Delta H \approx -12\,kJ\,mol^{-1}$ and $\Delta S \approx 46\,J\,mol^{-1}\,K^{-1}$ (each ±10–20%).

Clearly the notion of ion pairs is well established (Szwarc, 1968, 1972). In any but very polar solvents, it should therefore be assumed that ions invoked in mechanisms may exist partly or chiefly as pairs, which will have an effect on the form of the rate law for a reaction (see Problem 2.2). Szwarc (1972, p. v) emphasizes the importance of ion pairing in mechanisms. The presence or absence of ion pairing may lead to very different rates, or even different courses of reaction. In a study by Dunn and Buncel (1989) of the ethanolysis of *p*-nitrophenyldiphenyl phosphinate (**5**), the kinetic behavior of different alkali metal ethoxides indicated that EtO$^-$M$^+$ *ion pairs are more reactive than the free* EtO$^-$ *ions*.

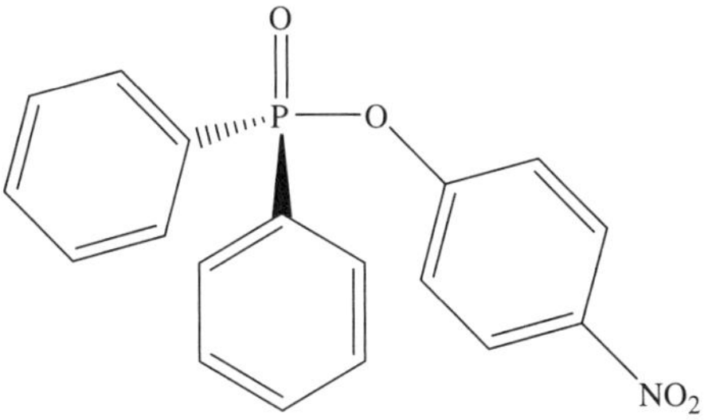

**5**

In more concentrated solutions, further aggregation to *ion triplets*, *quadruplets*, and so on, may occur (see, e.g., Erdey-Grúz, 1974, pp. 434–436; Fuoss and Accascina, 1959, pp. 249–272). This and the pairing of ions that have unequal charges make for mathematical complications, especially in the interpretation of the dependence of the conductance on concentration, and in kinetics. Balakrishnan *et al.* (2001) found that

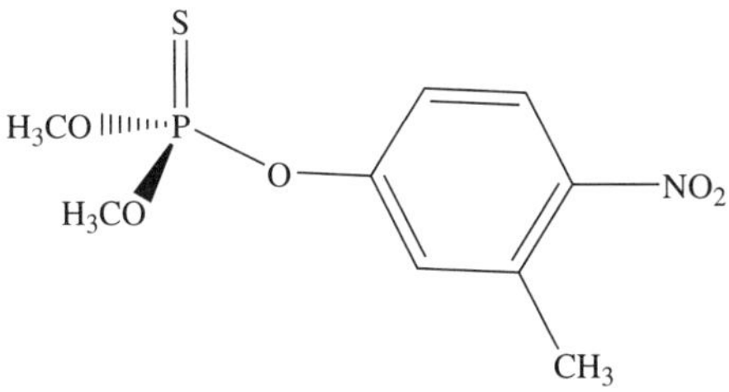

**6**

in degradation of the organophosphorus pesticide, fenitrothion (**6**), in ethanol in the presence of alkali metal ethanolate, one pathway was catalyzed by free ethanolate ion, but also by ion pairs and by $[2\,M^+.2EtO^-]$ quadruplets, the last being the most effective. Fuoss and Accascina (1959, p. 253) calculate that the concentration above which aggregation to higher clusters than pairs begins to be important is for a 1:1 electrolyte, $3.2 \times 10^{-7} \varepsilon_0^3$. For water at room temperature, this is about 0.155 M; for ammonia at $-33°C$, 1.8 mM.

In what may be seen as an extreme example of ionic association occurring in concentrated solution, during a study of the behavior of salt solutions in electrospray ionization mass spectrometry (Hao *et al.*, 2001) two distinct kinds of ionic species were detected, of mass/charge ratio corresponding to $(H_2O)_n Na^+$ and $(NaCl)_m Na^+$ (among others). In the early stages of evaporation of a charged droplet of a dilute aqueous NaCl solution *in vacuo* species of the first kind appear to be formed (Thomson and Iribarne, 1979); in later stages, the predominant species are of the second kind (Röllgen *et al.*, 1984). In the hydrated species, the abundance of successive species falls away in a fairly regular manner as the number of water molecules present increases. The salt clusters, however, favor certain values of $m$ ("magic numbers") that correspond to clusters of high symmetry (Doye and Wales, 1999), notably $m = 13$, which corresponds to a $3 \times 3 \times 3$ cubic cluster, containing 13 $Cl^-$ and 14 $Na^+$ ions.

When the same experiment was tried with dilute solutions of sodium carbonate in methanol (Hao and March, 2001), a surprising result was the appearance of species $(NaOCH_3)_n Na^+$. In studies on buffer solutions based on phosphate with initial pH values on either side of neutrality, it had been observed that the pH appeared always to shift away from seven as the droplet evaporated, judging by the species detected in the mass spectrum. The appearance of the species containing $NaOCH_3$ was, nevertheless, unexpected.

Davies (1962, chapter 9) gives a large table of the dissociation constants for a number of salts in a variety of solvents. He points out that the solvents appear to fall into two classes, which he calls *leveling* and *differentiating*. (These terms are also used in describing the effects of solvents on acids and bases; see Chapter 3.) The levelers are principally the hydroxylic solvents. In these solvents inorganic salts tend to be strong electrolytes, and where a comparison of the alkali metal salts of one acid is possible, the Li salt tends to be strongest and the Cs salt the weakest. The data for nitrates in methanol, $\varepsilon_r = 32.6$, are plotted against the reciprocal of the crystal radius of the cation in Figure 2.11. The probable reason for this order is that the smaller cations are able to bind solvent molecules more tightly, making a larger primary solvation sphere, so that Li(solv)$^+$ is effectively the largest and Cs(solv)$^+$ the smallest (Pregel *et al.*, 1995). All are, nevertheless, roughly comparable in size, except often lithium, which, being unique in having only a helium-type core, is able to bind four (usually) molecules exceptionally tightly.

A salt having two large ions of low charge, such as tetraethylammonium picrate (**7**), should show nearly perfectly the electrostatic effect, uncomplicated by the molecular nature of the solvent. Davies (1962, p. 97) shows data, claimed to be very precise, for the p$K$ of this salt in nitrobenzene, acetone, and 1,2-dichloroethane, plotted *versus* $1/\varepsilon$, giving a nearly perfect straight line, as required by the theory of Fuoss

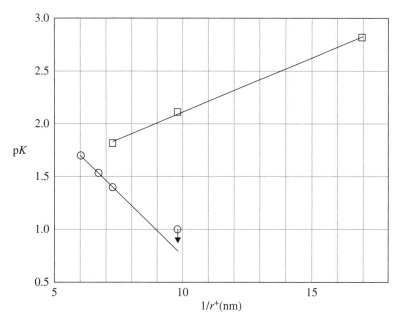

**FIGURE 2.11**   Ionic radius and ion-pair dissociation: $pK_{diss}$ for Cs, Rb, K, Na nitrates in methanol (circles) and K, Na, Li picrates in acetonitrile (squares) versus the reciprocal of the cation crystal radius. Data for $pK$ from Davies (1962, pp. 95, 96) and for radii from Atkins and de Paula (2010, p. 938). The value of $pK$ for $NaNO_3$ in methanol is reported as $<1$.

and Krauss, Equation 2.26. The tetraethylammonium halides and the alkali picrates in the same solvents give poorer graphs of this kind, because of the differing degrees of primary solvation of the small ions.

**7**

## 2.10   SOLVENT MIXTURES

Mixed solvents present both an opportunity and a problem. On the one hand, it is possible, by varying the proportions of components of a binary mixture, to vary certain properties continuously over a wide range. For example, water and

1,4-dioxane are miscible in all proportions. Mixtures can be prepared with any value of the relative permittivity from 2.2 to 78.5 at 25°C. Mixtures of acetonitrile ($\varepsilon_r = 36.2$) with dimethylformamide (36.7) are nearly *isodielectric*; other properties can be varied while keeping the relative permittivity nearly unchanged. The liquid range of a solvent can be extended slightly upward by admixture of a cosolvent with which the first forms a high-boiling azeotrope, for example, chloroform (b.p. 61.0°C) and acetone (56.3°C) form an azeotrope boiling at 63.4°C. Since adding an impurity to a liquid usually lowers its freezing point, the liquid range is extended downward, the eutectic temperature being below the freezing point of either component. For example, the eutectic mixture of LiCl (mp 613°C) and KCl (mp 776°C) melts at 355°C. Binary salt mixtures that are liquid at room temperature are available (Barthel and Gores, 1994).

On the other hand, it is not always clear what the environment of a solute species in a mixed solvent may be. Preferential solvation of ions by one component is common. Viscosity measurements of dilute salt solutions in water–methanol have been interpreted (Stairs, 1976, 1979) as showing free energies of preferential binding to various alkali halides of about $1 \, \text{kJ} \, \text{mol}^{-1}$ for water over methanol, without distinguishing between the cation and anion, and without detecting any trend. Strehlow and Schneider (1969) have distinguished cases in which the cation and anion are both preferentially solvated by the same component of a solvent mixture (*homoselective*) or by different components (*heteroselective*). Heteroselectivity can have interesting effects, such as enhancement of solubility of a salt in mixed solvent over that in either component, for instance, silver sulfate in water–acetonitrile. These solvents show partial miscibility below about $-1°C$. Addition of $0.1 \, \text{mol} \, l^{-1}$ of the homoselectively solvated salt $NaNO_3$ raises the critical solution temperature (defined for a binary system showing partial miscibility as the temperature above which the components are miscible in all proportions) nearly 3°, while the heteroselectively solvated salt $AgNO_3$ lowers it more than 6° (Strehlow and Schneider, 1971). Phase separation is favored by homoselective solvation; it is opposed by heteroselective solvation. Ràfols *et al.* (1997) have examined the effects of preferential solvation in a number of systems on $E_T(30)$, $\pi^*$, $\alpha$, and $\beta$.

Covington and Newman (1976) report the results of their NMR study of solutions of alkali halides in binary mixtures of water with seven other solvents, and UV–vis measurements on alkali metal bromides, iodides, and nitrates in water–acetonitrile, water–DMSO, and acetonitrile–methanol mixtures. Some interesting trends appear. For instance, in the nearly isodielectric binary solvent system water/hydrogen peroxide, the preferential solvation of the alkali metal ions by water over hydrogen peroxide decreases as the ionic radius increases; for the halide ions (as far as can be seen, since $Br^-$ and $I^-$ cannot be studied in the presence of peroxide) the trend is in the opposite direction. For water–methanol solutions, they find free energy differences favoring water at sites on $Na^+$, $Rb^+$, $Cs^+$, and $Cl^-$ of 1.28, 0.89, 0.66, and $0.96 \, \text{kJ} \, \text{mol}^{-1}$ and for $F^-$ of $-0.15 \, \text{kJ} \, \text{mol}^{-1}$ (i.e., favoring methanol).

A measure of selective solvation of single ions is the *isosolvation point*, that is, the solvent composition at which the mean numbers of molecules of the two solvent components bound to an ion are the same. Rahimi and Popov (1976) found by $^{109}Ag^+$ NMR that the isosolvation point for $Ag^+$ in water–acetonitrile was at 0.075 mol fraction of acetonitrile. This means that solvation by this component is preferred by about 6.3 kJ mol$^{-1}$ over water.

A complication in preferential solvation exists in the possibility of discriminatory effects (discussed further in Chapter 3) or their contrary. In the simple case represented by the equations:

$$MA_2 + B \rightleftharpoons MAB + A$$

$$MAB + B \rightleftharpoons MB_2 + A$$

if the composition of the solvent is at the isosolvation point, that is, the concentrations of $MA_2$ and $MB_2$ are equal, the fraction of M present as MAB at equilibrium should be ½. If, however, discrimination exists (in the sense, "You can't join my club unless you are like me") this fraction will be smaller. *Antidiscrimination* ("affirmative action"?) will make it larger.

Nevertheless, a number of authors report studies of solvent effects in mixed solvent systems. Amis and coworkers (1939, 1943), guided by Scatchard's (1932) theory of the effect of relative permittivity on rates of reactions between ions, constructed plots of the logarithm of the rate constant versus $1/\varepsilon_r$ for the reaction of tetrabromophenol–sulfonphthalein (**8**) with hydroxide in water–methanol and in water–ethanol, and for the reaction of ammonium ion with cyanate in water–methanol and in water–ethylene glycol. In each case they found that the curves were essentially linear in the water-rich region, down to relative permittivities of 55-45. As the amount of the organic component of the solvent increased further, curvature appeared, attributed to changes either in the mechanism or in the state of primary solvation of the reacting ions.

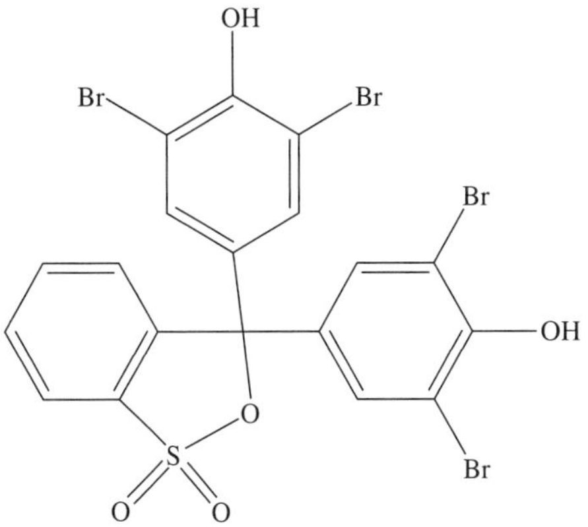

**8**

In highly structured solvents, such as water and alcohols, structural changes with composition may be manifested in physical properties. It has been suggested (Werblan *et al.*, 1971) that small additions of hydroxylic solvents to water enhance the special structure of water, up to a point. The $B$ coefficient of viscosity for solutions of alkali halides in water/methanol shows a minimum at $0.15-0.2$ mol fraction of methanol (Stairs, 1979), presumably because the structure-breaking effect of the ions finds the most structure to break in this range of solvent composition. The activation parameters $\Delta_{\ddagger}H$ and $\Delta_{\ddagger}S$ for the reaction of $Ni^{2+}$ with 2,2'-bipyridyl in water/ethanol have strong maxima at 0.08 mol fraction of ethanol (Tobe and Burgess, 1999). The effect largely cancels in $\Delta_{\ddagger}G$, as is often observed, but their plot of $\Delta_{\ddagger}H$ versus $\Delta_{\ddagger}S$ for the similar system with methanol as cosolvent shows curious kinks.

Solvolysis studies of alkyl halides in mixed solvents are common (see, e.g., Buncel and Millington, 1965). Grunwald and Winstein (1948) introduced the parameter $Y$, a measure of *ionizing power* based on the rate of solvolysis of 2-chloro-2-methylpropane (*t*-butyl chloride). It is a measure of the ability of the solvent to stabilize the ions that are generated in the unimolecular, rate-determining first step:

$$\left(CH_3\right)_3 CCl \rightarrow \left(CH_3\right)_3 C^+ + Cl^- \; : \; k_1 \left(slow\right)$$

$$\left(CH_3\right)_3 C^+ + 2HX \rightarrow \left(CH_3\right)_3 CX + H_2 X^+ \; : \; k_2 \left(fast\right)$$

HX represents an amphiprotic solvent. Since the first step is slow, $k=k_1$. $Y$ is defined by Equation 2.27:

$$Y = \log\left(\frac{k_S}{k_0}\right) \tag{2.27}$$

where $k_S$ is the rate constant for solvolysis in solvent **S**, and $k_0$ is the constant for the reference solvent, which is ethanol/water 80:20 v/v. Fainberg and Winstein (1956) report values of $Y$ for a number of aqueous–organic and a few organic–organic mixed solvent systems. In water–ethanol mixtures, $Y$ is not a linear function of the recorded volume percent, but is nearly linear on a mole percent scale.

The solvatochromic parameter $E_T(30)$, introduced by Reichardt and coworkers (Dimroth *et al.*, 1963), is the energy of a transition observed in the UV–visible absorption spectrum of the substituted pyridinium-*N*-phenoxide dye (**9**), 30th of a number tried. It has been measured in a great many single and mixed solvents. Cerón-Carrasco *et al.* (2014a, b) report values for 84 new solvents, and revised values for 184 previously measured. Reichardt (1994) provides a list of references. He points out that the dependence of this and other solvatochromic shifts on composition of binary solvent mixtures is not simple, being linear neither in mole fraction nor in volume fraction.

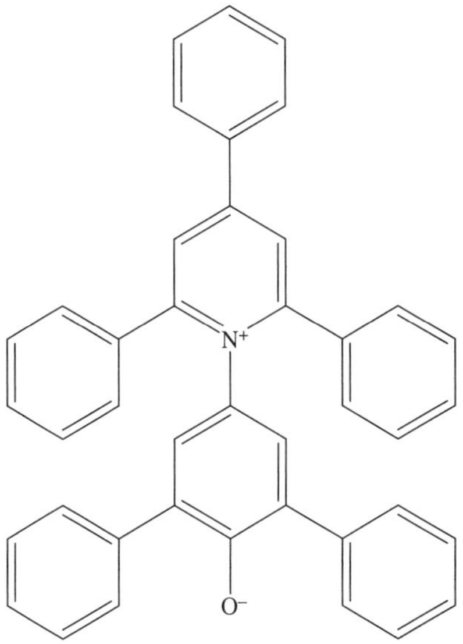

**9**

Brooker *et al.* (1965) report measurements of $\chi_R$ and $\chi_B$ (solvent polarity measures based on $\pi \rightarrow \pi^*$ transitions of two meropolymethine dyes, **10** and **11**) for a number of mixtures among water, methanol, 2,6-lutidine, 1,4-dioxane, and isooctane, which on volume fraction scales required cubic polynomials to be approximately fitted (by the present authors). In some instances, parameters such as $E_T(30)$ pass through a maximum. This can occur if the two components can, through hydrogen bonding, form a complex that is more polar than either alone, such as in DMSO–alcohol mixtures (Maksimović *et al.*, 1974). These effects, arising as they do from preferential solvation of the solvatochromic dyes, may not reflect the effects of varying solvent composition on reactions, unless the preferential solvation of reactants and products or activated complexes parallels the preferential solvation of the ground and excited states of the indicator dyes. It is apparent that conclusions drawn from effects on rates or equilibrium of solvent composition in mixed solvents cannot be confidently attributed, for instance, to changes in the bulk relative permittivity, or to any other single property of the mixed solvent.

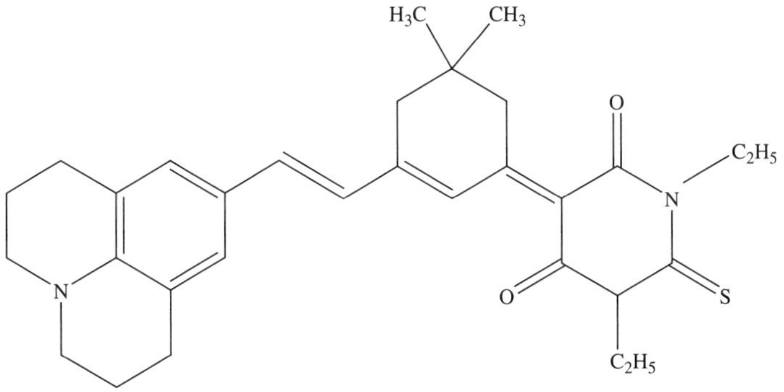

**10**

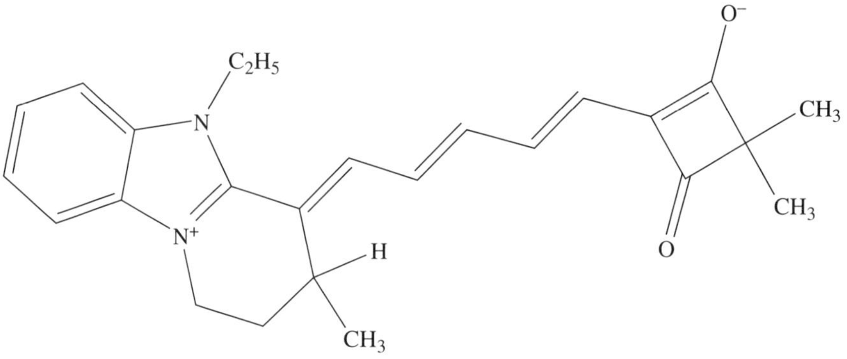

**11**

Langhals (1982a, b) fitted values of $E_T(30)$ for a number of binary mixtures by Equation 2.27, using a nonlinear least-squares procedure.

$$E_T(30) = E_D \ln\left(\frac{c_p}{c^*}+1\right) + E_T^0(30) \tag{2.28}$$

Here $E_T(30)$ represents the polarity of the binary mixture, and $E_T^0(30)$ that of the pure, less polar component; $c_p$ is the molar concentration of the more polar component, and $c^*$ and $E_D$ are adjustable parameters. Langhals showed that an equation of this form can also express the concentration dependence of several other measures of solvent polarity in binary mixtures. This equation predicts that, above a certain concentration, $E_T(30)$ should be linear in $\ln c_p$, as observed. Langhals also suggested (1982b) that it can be used in homologous series, by considering a linear alcohol $H(CH_2)_nOH$, for instance, as a mixture of 1 mol of the polar H–OH group with $n$ moles of the nonpolar $CH_2$. This means that $c_p$ is equal to the reciprocal of the molar volume of the alcohol, that is, $c_p = \rho/M$. Guided by this concept, he obtained a plot of $E_T(30)$ linear in $\ln c_p$, or rather one with two distinct linear sections.

The present authors note that this same concept, translated into mole fraction terms, gives for the mole fraction of the polar group, $x_p = 1/(1+n)$, and that of the nonpolar chain segments, $x_0 = n/(1+n)$. If the effect is assumed to be additive, Equation 2.28 results:

$$E_T(30) = \frac{E_p}{1+n} + \frac{nE_0}{1+n} \tag{2.29}$$

Here $E_p$ and $E_0$ should represent the values of $E_T(30)$ appropriate to the polar group and the $CH_2$ group, respectively. Figure 2.12 is Reichardt's (1994, figure 1), with the addition of the curve calculated from Equation 2.29, with $E_p = 63.37$ and $E_0 = 46.27$.

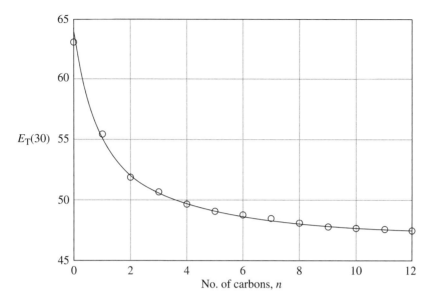

**FIGURE 2.12** $E_T(30)$ values for water and normal alcohols versus number of carbons. The curve is represented by $E_T(30)=(E_p+nE_0)/(n+1)$, but the data were fitted using the linearized form, Equation 2.29. The polar group contribution $E_p$, $=63.37$; the methylene group contribution $E_0=46.27$.

These values may be compared with the values for water (63.1) and for a normal alkane (31.0−31.1). A similar fit of data for nitriles (three points only) gave $E_p=51.8$ and $E_0=39.4$. This latter value of $E_p$ may be taken as an estimate of the value of $E_T(30)$ for HCN, not otherwise known. Why the $E_0$ values are consistently higher than the alkane values is not clear. Cerón-Carrasco *et al.* (2014b) extended this treatment to other homologous series.

For graphical or least-squares determination of the constants $E_p$ and $E_0$, Equation 2.28 may be linearized in the form:

$$\left(1+n\right)E_T\left(30\right)=E_p+E_0n \tag{2.30}$$

A plot of $(1+n)\ E_T(30)$ versus $n$ for these straight-chain alcohols was linear from $n=0$–$12$, with a correlation coefficient $r=0.999$. Secondary, tertiary, and branched-chain alcohols all fell below the line, showing that, in general, correlations between properties are greatly dependent on structure.

## 2.11 SALT EFFECTS

A reaction medium consisting of a pure solvent plus added salt is a mixed solvent of a special kind. Reactions involving no ions or only one, among reactants or products, or in the transition state, should not be subject to large salt effects. Where both reactants in a bimolecular process are charged, however, electrostatic contributions to the energy of activation should exist. These contributions are, according to the Debye–Hückel theory, reduced by the presence of other ions.

The magnitude of the kinetic salt effect for a bimolecular reaction depends strongly on the charges on the reacting species. Equation 2.30, derived from the Debye–Hückel theory, has been shown to represent most experimental data well (Atkins, 1998, p. 836).

$$\log\left(\frac{k}{k_0}\right) = 2Az_+z_-I^{1/2} \tag{2.31}$$

Here $A$ is the constant in the Debye–Hückel limiting law and $I$ is the ionic strength, defined earlier. For water at 25°C, $A = 0.51$.

Following Pethybridge and Prue (1972), a plot of the logarithm of the rate constant against $I^{1/2}/(1+I^{1/2})$, a better abscissa than simply $I^{1/2}$ (cf. the Davies approximation, Equation 2.23), should yield a straight line of slope equal to $2A$ times the product of the charges of the reacting ions, $z_+z_-$. Tobe and Burgess (1999, p. 359) show such plots for a variety of reactions of mixed character, organic and inorganic. Figure 2.13 shows a plot of the slopes of the lines in their figure versus the charge product, showing the trend clearly, with nearly the correct slope, 1.05. Tobe and Burgess caution,

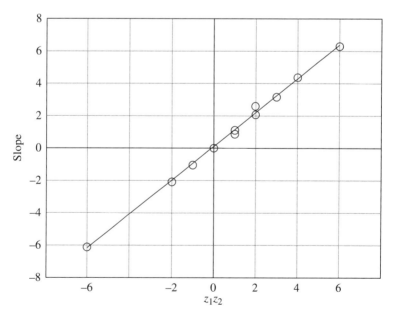

**FIGURE 2.13**   Kinetic salt effect; slope of the graph of $\log(k/k_0)$ vs. $I^{1/2}/(1+I^{1/2})$, where $I^{1/2}$ is the ionic strength, for 11 reactions between ions of different charge type in aqueous solution at 25°C versus the product of the charges on the reacting ions. Data from Tobe and Burgess (1999, p. 359).

however, that specific salt effects can occur that may obscure the electrostatic effect, especially where organic ions are present.

## PROBLEMS

**2.1**  In dilute solution in 1,4-dioxane, the rational activity coefficients (as logarithms) of the four participants in the reaction in Problem 1.2 are estimated as:

|         | Ethanol | Acetic acid | Ethyl acetate | Water |
|---------|---------|-------------|---------------|-------|
| $\ln(f)$ | 1.02    | 1.39        | 0.08          | 2.1   |

Calculate the practical equilibrium constant of the same reaction in dioxane.

**2.2**  Using any form of the Debye–Hückel equation, estimate how dilute a solution of sodium chloride must be for the solubility of barium sulfate in this solution at 25°C to be within 5% of the value calculated from the solubility-product constant.

**2.3**  Consider a reaction between ions $A^+$ and $B^-$ that proceeds through the ion pair $A^+B^-$, that is,

$$A^+ + B^- \rightleftharpoons A^+B^- ;\text{rapid equilibrium,}\quad K = \frac{\left[A^+B^-\right]}{\left[A^+\right]\left[B^-\right]}$$

$$A^+B^- \left( \rightarrow AB^\ddagger \right) \rightarrow \text{product; Slow, } k.$$

If $C$ is the analytical concentration of the salt AB, show that the general expression of the rate law is

$$\text{rate} = \frac{d[P]}{dt} = \left( \frac{k}{2K} \right)\left( (1+2KC) - (1+4KC)^{1/2} \right)$$

The order of the reaction with respect to $C$ is

$$\text{ord} = \frac{d\ln(\text{rate})}{d\ln C} = \frac{2KC\left(1 - (1+4KC)^{-1/2}\right)}{(1+2KC) - (1+4KC)^{1/2}}$$

Find the limiting values of the order at small and large values of $KC$.
At what value of $KC$ is the order 3/2?

**2.4**  Show by a statistical argument that in a system containing the species $MA_2$, MAB, and $MB_2$ at equilibrium, with equal concentrations of $MA_2$ and $MB_2$

(at the isosolvation point), the fraction of M present as MAB should be ½ in the absence of discrimination. Assuming this is true for the silver ion in water/acetonitrile and that the coordination number of $Ag^+$ is 2, calculate the equilibrium constant and the value of $\Delta G^\circ$ at 298 K for each of the following reactions, using the finding of Rahimi and Popov (1976) that the isosolvation point is at 0.075 mol fraction of acetonitrile.

$$Ag(H_2O)_2^+ + CH_3CN \rightleftharpoons Ag(H_2O)(CH_3CN)^+ + H_2O$$

$$Ag(H_2O)(CH_3CN)^+ + CH_3CN \rightleftharpoons Ag(CH_3CN)_2^+ + H_2O$$

# 3

# REACTIVE SOLVENTS

## 3.1 SPECIFIC SOLUTE/SOLVENT INTERACTIONS

Negative deviations from Raoult's law imply that the interaction energy of two unlike molecules is greater than the mean of the two "like–like" interactions. (We have seen that mere difference between the molecules leads to the opposite effect.) Specific interactions that could cause negative deviations include hydrogen bonding, Lewis acid–base reaction, and charge transfer.

*Solvation* and *solvolysis* must be distinguished. The former term implies that the solute remains intact, though with one or more attached solvent molecules. The strength of attachment may vary widely and may involve only van der Waals interactions as described in Section 2.1 but may include specific interactions: through hydrogen bonding (see Section 3.2); through the formation of covalent bonds by the donation and acceptance of pairs of electrons, usually discussed under the heading of Lewis acid–base chemistry (Section 3.7); or through charge transfer. This last we will not discuss further than to note the familiar example of molecular iodine, which in the vapor and in solution in solvents such as tetrachloromethane is violet in color but in benzene, for instance, is brown, owing to the ability of benzene to receive electron density, transferred from a nonbonding orbital of $I_2$ into one of its antibonding $\pi^*$ molecular orbitals.

*Solvolysis*, on the other hand, describes a reaction between the solute and the solvent, in which the identity of the solute is lost. In sulfuric acid, typical of strongly acidic solvents, most salts, if they are at all soluble, react to produce the conjugate acid of the anion and the sulfate of the cation (which may precipitate), for example,

$$2 \ Fe(NO_3)_3 + 3 \ H_2SO_4 \rightarrow Fe_2(SO_4)_3(s) + 6 \ HNO_3$$

*Solvent Effects in Chemistry*, Second Edition. Erwin Buncel and Robert A. Stairs.
© 2016 John Wiley & Sons, Inc. Published 2016 by John Wiley & Sons, Inc.

In aqueous solution, many salts are to different extents solvolyzed (hydrolyzed), the solution becoming acidic if the cation is small or highly charged:

$$Mg^{2+} + 7H_2O \rightarrow [Mg(H_2O)_6]^{2+} + H_2O \rightleftharpoons [Mg(H_2O)_5OH]^+ + H_3O^+$$

or basic if the anion is derived from a weak acid:

$$CH_3COO^- + H_2O \rightleftharpoons CH_3COOH + HO^-$$

An example of solvolysis important in physical organic chemistry is the solvolysis of *t*-butyl chloride (2-chloro-2-methylpropane), the rate of which is the basis of the Grunwald and Winstein (1948) parameter *Y*, described in Section 2.10.

## 3.2 HYDROGEN BONDING

Species in which a hydrogen atom is attached to an atom that is intrinsically electronegative, such as oxygen, or that is made electronegative by other attached atoms, as is the carbon atom in chloroform, possess a local (bond) electrical dipole, with hydrogen bearing a partial positive charge, owing to withdrawal of electron density toward the electronegative atom. This also results in diminution of the van der Waals radius of the hydrogen, so it is more able to approach a center of negative charge in the form of an unshared electron pair on another molecule. The electrostatic attraction is accompanied by some orbital involvement. The resulting attractive interaction is termed a *hydrogen bond* (H bond), and the molecule bearing this hydrogen is termed a *hydrogen bond donor* (HBD). A *hydrogen bond acceptor* (HBA) is a molecule that possesses a site where an unshared pair of electrons on an atom bearing a partial negative charge is sterically accessible to the approaching HBD molecule. Intramolecular hydrogen bonds are also possible. Typical hydrogen bonds, for instance, those between water molecules and between the OH groups in alcohols, which have both HBA and HBD characters, have strengths in the region $15–20\,kJ\,mol^{-1}$. Hydrogen bonds in amines are weaker, while those involving fluorine (for instance, in the ion FHF⁻) are stronger.

The effect of H bonding on the vapor–liquid equilibrium is expected to depend both on the strength and number of H bonds per molecule. Water has two donor sites and two acceptor sites for H bonding, tetrahedrally oriented, so the formation of three-dimensional structures is possible. The high melting and freezing points of water relative to other molecules of comparable size, such as methane, are obvious consequences of this ability of water to form a network in three dimensions. Ammonia has only one acceptor site, while hydrogen fluoride and methanol each have only one donor site. These can form chains with limited branching. The relative strengths of H bonds are in the order NHN < OHO < FHF.

The order of the boiling points $NH_3 < HF < H_2O$ shows that the number of possible bonds is more important than their strength.

The structure and solvent properties of water are thus dominated by hydrogen bonding. In ice, the network is essentially complete; in the liquid, it is imperfect but persists to temperatures above the melting point and is responsible for the decrease in density of water as the freezing point is approached. A large part of water's solvent power for salts derives from its ability to form hydrogen bonds with anions. Its miscibility with 1,4-dioxane and with acetone results from hydrogen bonding with the oxygens in these HBA molecules.

Hydrogen atoms in hydrocarbons are essentially unable to take part in hydrogen bonding, for the electronegativities of carbon (2.5) and hydrogen (2.1) are not sufficiently different, but in compounds in which a carbon bears electronegative substituents, or is conjugatively influenced by electron-withdrawing groups, the situation may change. Chloroform is a weak HBD substance and acetone is a moderately strong HBA. Binary mixtures of these exhibit negative deviations from Raoult's law, sufficient to cause a maximum in the boiling point (azeotrope) at about 0.8 mol fraction of chloroform.

Hydrogen bonding may be viewed as a sort of tentative acid–base interaction. If an HBD molecule made an outright gift of a proton, instead of just sharing it, we should call it an acid. The receiving molecule would then be acting as a base. In the next sections, we take up the consequences of this and consider the Brønsted–Lowry and Lewis views of acids and bases.

## 3.3 ACIDS AND BASES IN SOLVENTS

In any discussion of acids and bases in solvents, it is necessary to distinguish between solvents that are themselves capable of acting as acids or bases and those that cannot do so. In the context of the Brønsted–Lowry definition, molecules that contain no hydrogen clearly cannot act as acids, but it is difficult to conceive a molecule that could in no circumstances be protonated. For example, the species $CH_5^+$ is well known from mass-spectrometric studies, and protonation of alkanes has been demonstrated in superacid solutions (Fabre *et al.*, 1982; Olah *et al.*, 1985). Similarly in the Lewis context, it is difficult to conceive a solvent that could in no circumstances become involved in electron-pair sharing either as donor or as recipient. Clearly, though, solvents that are too weakly acidic or basic to interact appreciably with any given set of Brønsted or Lewis acids and bases *are* conceivable. They are best called "inactive" or "inert," bearing in mind that these are relative terms. Aprotic or inert solvents do not require us to add anything to the considerations of the previous chapter, as they function as mere media. *Active solvents*, on the other hand (the term may as well include proton or electron-pair donor or acceptor solvents), are able to participate in reactions, either as reactants or products, or, more subtly, as catalysts. We shall consider proton donor and acceptor solvents first.

## 3.4   BRØNSTED–LOWRY ACIDS AND BASES

In this scheme, an acid is any species (not "substance") that can donate a proton to another species, and a base is a species that can receive a proton (see, e.g., Stewart, 1985). Since bare protons cannot exist for any appreciable time in a liquid, acids and bases always react in mutual fashion by proton transfer:

$$HA + B \rightleftharpoons A + HB$$

Charges have been omitted from this equation; A and B will always be one unit more negatively charged than HA and HB. HA and A are termed *conjugate* acid and base. Either HA or B may represent a solvent molecule. *Strength* of an acid is relative; if HA is a stronger acid than HB, the equilibrium will lie to the right. This is exactly equivalent to saying that B is a stronger base than A.

Solvents that are *amphiprotic*, that is, those that can both give and receive a proton, are capable of *autoprotolysis* or *autoionization*, for example, for water:

$$2H_2O \rightleftharpoons H_3O^+ + HO^-$$

or, more generally, if we represent the solvent molecule by HS:

$$2HS \rightleftharpoons H_2S^+ + S^-$$

In the pure solvent, the two ions will be present in equal concentrations, but the presence of a foreign acid or base is signaled by the presence in solution of an excess of one or the other of the characteristic ions of the solvent. This is the basis of the Arrhenius definition of an acid as a substance (not "species"), which when dissolved in water yields "hydrogen ions" (hydrated $H_3O^+$ ions). An acid substance is "strong" in a given solvent if it can donate protons to the solvent to such an extent that a $c$ molar solution of the substance in the solvent yields a concentration of the *solvo-acid* ion, $[H_2S^+] = c \, mol\,l^{-1}$, approximately. (Here, we have used the convention that the square brackets [ ] represent the actual concentration of a species, while $c$ represents the formal concentration of a substance.) If $[H_2S^+] \ll c$, the acid is "weak" in that solvent. Since $[H_2S^+]$ can never exceed $c$, the strongest acids all appear equally strong. This is called *leveling*.[1] *Differentiating* solvents, on the other hand, are those that are weak enough bases so that the reaction of most acids with the solvent is incomplete, and different acids react to differing degrees. In water, hydrogen chloride and acetic acid are differentiated, but in liquid ammonia, they are leveled; both are strong.

The activities (or, approximately, concentrations) of the solvo-acid and *solvo-base* (solvent minus a proton) are linked through the autoprotolysis or *ion-product constant*:

$$K_{ip} = [H_2S^+][S^-]$$

---

[1]Leveling up, that is, like the Portuguese revolution of 1910, which made everyone "Your Excellency," in contrast to the French one, which leveled everyone down to "Citizen" (Maurois, 1927).

**TABLE 3.1 Autoionization Constants of Selected Solvents**[a]

| Solvent | $t/°C$ | $pK_{ip}$ | $pH_n$ | Solvent | $t/°C$ | $pK_{ip}$ | $pH_n$ |
|---|---|---|---|---|---|---|---|
| Water | 25 | 14.0 | 7.0 | Hydrogen fluoride | 0 | 13.7 | 6.9 |
| Ammonia | –50 | 33 | 16.5 | Sulfuric acid[b] | 25 | 3.57 | 1.8 |
| ,, | –33 | 27 | 13.5 | Phosphoric acid | 25 | 0.9 | 0.5 |
| 1,2-Diaminoethane | 25 | 15.3 | 7.7 | Acetic acid | 25 | 14.5 | 7.2 |
| Methanol | 25 | 16.7 | 8.4 | Formic acid | 25 | 6.2 | 3.1 |
| Ethanol | 25 | 18.9 | 9.5 | Fluorosulfonic acid | 25 | 7.4 | 3.7 |

[a] Data largely from Jolly(1970, p. 99); see also Fabre *et al.* (1982).
[b] The autoionization of sulfuric acid is complicated by the existence of two modes: ionic dehydration ($2H_2SO_4 \rightleftharpoons H_3O^+ + HS_2O_7^-$; $K_{id} = 5.1 \times 10^{-5}\,mol^2\,kg^{-2}$) and autoprotolysis ($2H_2SO_4 \rightleftharpoons H_3SO_4^+ + HSO_4^-$; $K_{ap} = 2.7 \times 10^{-4}$) (Cerfontain, 1968; Gillespie and Robinson, 1965).

which for water at about 25°C is $1.00 \times 10^{-14}\,mol^2\,l^{-2}$. Some values for other solvents are listed in Table 3.1. Since the concentrations of both ions must be equal in the pure solvent, each must be equal to $K_{ip}^{1/2}$, and the "neutral point" on a pH scale for any solvent is $pH_n = -\frac{1}{2}\log_{10}K_{ip}$: about 7.00 for water at 25°C, about 15 for ammonia at –33°C, and about 7.2 for acetic acid at 25°. If the pH is greater than $pH_n$, protons have been transferred from the solvent to the solute; the solution is basic. If the pH is less than $pH_n$, protons have been transferred to the solvent; the solution is acidic. The pH scales are thus different in each solvent. So are the related "relative acidity index" scales, $A = pH_n - pH$, which place the neutral point for each solvent at zero, and do away with the confusing contrary motion of acidity and pH (Stairs, 1978a, 1983).

The degree of self-ionization depends on both the acid strength and the basic strength of the molecule. For example, for water (assuming molar concentrations approximate to activities; $H^+$ is in some standard state, perhaps in the gas phase),

$$H_2O \rightleftharpoons H^+ + HO^-; K_a$$

$$H_2O + H^+ \rightleftharpoons H_3O^+; K_b$$

$$2H_2O \rightleftharpoons H_3O^+ + HO^-; K = K_b K_a = [H_3O^+][HO^-]/[H_2O]^2 = 3.2 \times 10^{-18}.$$

(The conventional $K_{ip}$ divided by the molar concentration of $H_2O$ in pure water, squared.)

Similarly for ammonia at 20°C,[2]

$$2NH_3 \rightleftharpoons NH_4^+ + H_2N^-; K = [NH_4^+][H_2N^-]/[NH_3]^2 = \sim 1.5 \times 10^{-17}.$$

and for hydrogen fluoride,

$$3HF \rightleftharpoons FH_2^+ + F_2H^-; K = [FH_2^+][F_2H^-]/[HF]^3 = \sim 1.4 \times 10^{-19}.$$

[2] Kilpatrick and Jones (1967, p. 59) give $K_{ip} = 2.07E-11$ at 0°C. Fabre *et al.* (1982, p. 593) give $pK_{ip} = 13.7$, so $K_{ip} = 2E-14$, temperature not stated, perhaps 20°C. We used the latter value.

## 3.5  ACIDITY FUNCTIONS

Following the pioneering work of Hammett and coworkers (1932, 1970), in establishing the acidity functions, $H_0$, $H_+$, $H_-$, and so on, it became apparent that organic bases of different structures lead to different scales. Efforts at reconciling them date back at least to Bunnett and Olsen (1966). Cox and Yates (1983) list 39 different symbols, representing 425 more-or-less distinct scales, depending on different classes of indicator bases and suitable for use in different solvent systems and acidity/basicity ranges.

It would, of course, be desirable to define a single scale that measures proton activity, $a_{H^+}$, such that in dilute aqueous solution, it reduces to the accepted scale: $pH = -\log_{10}\left(a_{H^+}\right)$. Writing for a particular base A in acid solution the equilibrium

$$AH^+(solv) \rightleftharpoons A(solv) + H^+(solv)$$

the corresponding equilibrium constant is

$$K_{AH^+} = \frac{a_{H^+} a_A}{a_{AH^+}} = \frac{a_{H^+}[A]}{[AH^+]} \cdot \frac{\gamma_A}{\gamma_{AH^+}} \tag{3.1}$$

or, in logarithmic form,

$$pK_{AH^+} = -\log a_{H^+} + \log I_A - \log\left(\frac{\gamma_A}{\gamma_{AH^+}}\right) \tag{3.2}$$

where the ionization ratio $I_A = [AH^+]/[A]$. Rearranging

$$-\log(a_{H^+}) = pK_{AH^+} - \log I_A + \log\left(\frac{\gamma_A}{\gamma_{AH^+}}\right) \tag{3.3}$$

it is apparent that the problem of determining $-\log a_{H^+}$, once the ionization ratio and $pK_{AH^+}$ are known, lies in the behavior of the last term, containing the activity coefficients. Hammett's method depends on the assumption that, for a series of progressively weaker bases used to ascend the scale by overlapping their ranges of measurable ionization ratios, the activity coefficient ratios for two bases A and B in the same solution are equal. (Cox and Yates (1983) describe this assumption as the "zero-order approximation.") For the second base, B,

$$-\log\left(a_{H^+}\right) = pK_{BH^+} - \log I_B + \log\left(\frac{\gamma_B}{\gamma_{BH^+}}\right) \tag{3.4}$$

and, by subtraction,

$$pK_{AH^+} - pK_{BH^+} - \log I_A + \log I_B = \log\left(\frac{\gamma_A \gamma_{BH^+}}{\gamma_B \gamma_{AH^+}}\right) = 0 \qquad (3.5)$$

so $pK_{BH^+}$ can be determined, and the scale continued as far as the ionization ratio of B can be measured. Hammett's $H_0$ scale depends on the use of a series of aromatic amines, of progressively weaker basicity. He defines

$$H_0 = pK_{AH^+} - \log I_A = -\log[H^+] - \log\left(\frac{\gamma_{H^+} \gamma_A}{\gamma_{AH^+}}\right) \qquad (3.6)$$

To extend this function to higher levels of acidity requires a succession of suitable weak bases. Each should have a $pK_{AH^+}$ value close enough to that of its predecessor that an overlapping range of acidity exists wherein the ratios $I_A$ for both can be measured spectroscopically. It has been determined over a large range of acidities, from dilute aqueous solutions into the superacid range.

The foregoing assumes that A and B are of similar structure. If they are not, the assumption fails. Consider A and B as representing bases of two different structural classes. Most obviously, if the charges on the bases are different, the interionic parts of the activity coefficient ratios will differ. Using Davies's Equation 2.23 as a rough guide, if the interionic part of the activity coefficient of a univalent ion is given by $\gamma_1^e$, then $\log(\gamma_z^e) = z^2 \log(\gamma_1^e)$, and

$$-\log \Gamma_z^e = -\log\left(\frac{\gamma_{H^+}^e \gamma_{A^z}^e}{\gamma_{AH^{z+1}}^e}\right) \approx 2z \log(\gamma_1^e) \qquad (3.7)$$

where $z$ is the charge on the base, corresponding to the subscript on $H_0$, $H_+$, or $H_-$.

Even if A and B bear the same charge but are of different structures (e.g., a primary aromatic amine and an amide), the interactions of the bases and their protonated forms with solvent will differ, and the assumption still fails. Let A and B represent classes of bases of different structures. A scale dependent on bases of the B class can be established, but it will be different from the scale derived from class A, owing to differences in the mode and strength of solvation of the protonated and unprotonated species belonging to the A and B classes. Hammett (1970, p. 274) suggests, "There is no reason except the large accumulation of data which accompanies its seniority in time for preferring $H_0$ as a measure of acidity over any of the other acidity functions." This is still true and must remain so until some means exists of determining the degree of solvation of the protonated and unprotonated forms of each of the indicator bases, perhaps by theoretical calculations of the sort described in Chapter 5.

To reconcile the various scales, several "first-order" approximations have been proposed. Bunnett and Olsen (1966, and references in Cox, 2000) assume that the logarithm of the activity coefficient ratio for a base of class B is not equal to that for a Hammett base, A, but is proportional to it, that is,

$$\log \Gamma_B = \log \left( \frac{\gamma_{H^+} \gamma_B}{\gamma_{BH^+}} \right) = (1 - \varphi) \log \left( \frac{\gamma_{H^+} \gamma_A}{\gamma_{AH^+}} \right) = (1 - \varphi) \log \Gamma_A \tag{3.8}$$

So

$$H_B + \log[H^+] = -\log \Gamma_B = -(1 - \varphi) \log \Gamma_A = (1 - \varphi)(H_0 + \log[H^+]) \tag{3.9}$$

Substituting in Equation 3.9 the definition $H_B = pK_{BH^+} - \log I_B$, analogous to Equation 3.6, we obtain

$$\log I_B + H_0 = pK_{BH^+} + \varphi(\log[H^+] + H_0) \tag{3.10}$$

and the B scale can be related to the A scale ($H_0$) by plotting $\log I_B + H_0$ versus $\log[H^+] + H_0$, for by Equation 3.10, this plot gives the values of $pK_{B^+}$ from the intercept and $\varphi$ from the slope.

An alternative method of reconciling the various scales is the *excess acidity method*, developed by Marziano *et al.* (1973, 1977) and confirmed and extended by Cox and Yates (1981, 1983; Cox, 1987, 2000). Cox defines $X$ as the excess acidity over the stoichiometric acid concentration, writing

$$\log I - \log[H^+] = m^* X + pK_{BH^+} \tag{3.11}$$

for any indicator base B. Here, $m^*$, the slope parameter, corresponds to $1 - \varphi_e$. The data for many bases in aqueous HCl (<40%), $HClO_4$ (<80%), and $H_2SO_4$ (<99.5%) have been correlated by computer (Cox and Yates, 1978), and the results summarized by two empirical equations expressed in terms of weight percent of acid, which we have translated to mole fraction $x_2$, in the form of Equation 3.12. Table 3.2 lists the coefficients found by least squares fit to conform to Cox and Yates's formulation:

$$X = c_0 + c_1 x_2 + c_2 x_2^2 + \cdots \tag{3.12}$$

Figure 3.1 shows the excess acidity of aqueous solutions of sulfuric, perchloric, and hydrochloric acids calculated from the original equations, plotted against the mole fraction of each acid. The greater intrinsic strength of perchloric acid is apparent at mole fractions above 0.1.

**TABLE 3.2   Coefficients in the Equation $X = c_0 + c_1 x_2 + c_0 x_2^2 + \cdots$ Representing the Excess Acidity in Aqueous Solutions of Sulfuric, Perchloric, and Hydrochloric Acids As Functions of the Mole Fraction of Acid, $X_2$, with the Standard Error of Estimate, $S^a$**

|         | Sulfuric          | Perchloric         | Hydrochloric       |
|---------|-------------------|--------------------|--------------------|
| $c_0$   | $-0.02 \pm 0.01$  | $-0.016 \pm 0.008$ | $-0.013 \pm 0.002$ |
| $c_1$   | $13.30 \pm 0.20$  | $10.42 \pm 0.31$   | $11.92 \pm 0.11$   |
| $c_2$   | $17.51 \pm 1.03$  | $90.7 \pm 3.4$     | $40.56 \pm 1.02$   |
| $c_3$   | $-45.92 \pm 1.84$ | $-238.5 \pm 13.2$  | $-101.78 \pm 2.75$ |
| $c_4$   | $26.61 \pm 1.05$  | $179.4 \pm 16.5$   | —                  |
| $s$     | $0.04$            | $0.02$             | $0.01$             |

$^a$ Recalculated after Cox and Yates (1978).

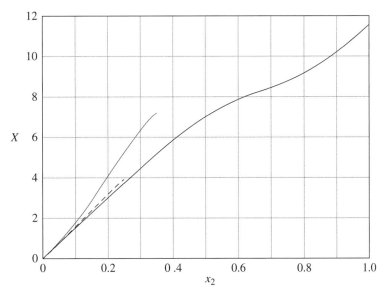

**FIGURE 3.1**   Excess acidity, X, in aqueous perchloric (light curve), sulfuric (black), and hydrochloric (dashed) acid solutions versus mole fraction of the acid, $x_2$. Source: Cox (2000), with permission.

## 3.6   ACIDS AND BASES IN KINETICS

### 3.6.1   Mechanisms

Cox (2000) shows how the concept of excess acidity can be used in kinetic studies as an aid in assigning mechanisms. For example, data (Bell *et al.*, 1956; Cox, 2000) for the acid-catalyzed depolymerization of the cyclic trimer of formaldehyde (trioxane) in several media (aqueous sulfuric, perchloric, and hydrochloric acids) were plotted as $\log k\psi - \log C_{H^+}$ versus $X$ ($k\psi$ represents the observed pseudo-first-order rate constant). The plots were highly linear over two orders of magnitude of $X$, with a slope of $1.091 \pm 0.010$. This was shown to be consistent with the *A-1 mechanism*:

$$S + H^+ \underset{\text{fast}}{\overset{K_{SH^+}}{\rightleftharpoons}} SH^+ \xrightarrow[\text{slow}]{k_1} \text{products}$$

Here, S represents trioxane (**1**) and $K_{SH^+}$ is the reciprocal of the (large) acid dissociation constant of the protonated trioxane (**2**).

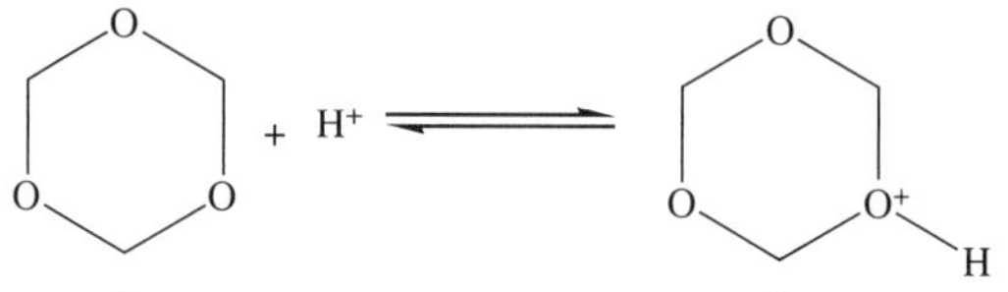

**1**              **2**

In contrast, in the *A-2 mechanism*, the species $SH^+$ requires the assistance of a nucleophile, usually from the solvent, in the rate-determining step:

$$S + H^+ \underset{\text{fast}}{\overset{K_{SH^+}}{\rightleftharpoons}} SH^+ + Nu: \xrightarrow[\text{slow}]{k_2} \text{products}$$

An example is the enolization of acetone in aqueous acid:

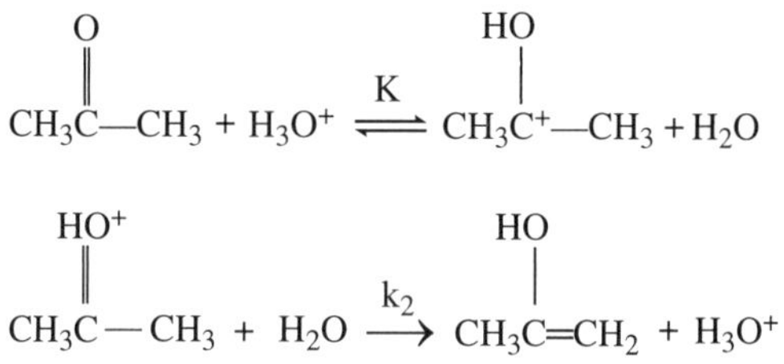

The A-2 mechanism can be distinguished from A-1 by curvature in the $\log k\psi - \log C_{H^+}$ versus $X$ plot (Cox, 2000).

### 3.6.2 General Acid–Base Catalysis and Brønsted Relationships

Reference was made above to involvement of the solvent in a reaction as catalyst. In a protic solvent, reaction may be catalyzed by the solvonium ion only (the hydronium ion in water). This is specific hydrogen-ion catalysis. On the other hand, the reaction may be catalyzed by any acidic species present in the solution (general acid catalysis). The solvent molecule itself may be a catalyst. Base catalysis, similarly, may be specific or general. For instance, the apparent rate constant for a reaction in aqueous solution might be represented by

$$k_{obs} = k_0 + k_w[H_2O] + k_{H^+}[H_3O^+] + k_{HO^-}[HO^-] + k_a[HOAc] + k_b[AcO^-] \quad (3.13)$$

where the first term represents the "uncatalyzed" rate constant, the second the constant for catalysis by water acting as acid or as base, and the other terms as suggested by the subscripts. Disentangling these contributions requires rate measurements over ranges of pH with different concentrations of buffer components (Bell, 1941, 1973; Jencks, 1985).

Brønsted and Pederson (1924) showed that for a series of acids of similar nature, for example, carboxylic acids, their catalytic constants $k_a$ for a particular reaction are related to their dissociation constants $K_a$ according to Equation 3.14 (a or b), called the Brønsted relation:

$$k_a = GK_a^{\alpha} \quad (3.14a)$$

or

$$\log k_a = g + \alpha \log K_a \quad (3.14b)$$

The constants $G$ or $g$ and $\alpha$ depend on the reaction, the solvent, and the temperature. Multiplied by $2.303RT$, each logarithm in Equation 3.14b becomes a Gibbs energy. This is probably the earliest example of a linear free energy relationship, of which more will follow in Chapter 4. Bell and Higginson (1949) report on a study of the dehydration of acetaldehyde hydrate in acetone, catalyzed by 32 carboxylic acids and 15 phenols. As Figure 3.2 shows, a good linear plot was obtained, the acid dissociation constants covering a range of $10^9$ and the rate constants a range of $10^5$.

Similar relationships exist between the rate constant of any reaction in which the rate-determining step is proton transfer from an acid and the dissociation constant of the acid. A Brønsted-type plot of the logarithm of the rate constant for the reaction of each acid versus the corresponding dissociation constant gives a straight line of slope $\alpha$. There are theoretical reasons to expect the slope to be unity for very slow reactions and zero for very fast ones. In practice, one can access only a small part of the curve with a slope of intermediate value, and the curvature is hardly apparent. Values of $\alpha$ found commonly range from 0.3 to 0.9 (Wiberg, 1964, p. 401).

For reactions involving nucleophiles, Brønsted-type plots may be made, in which the logarithms of the rate constants of reactions between a series of nucleophiles and a single substrate are plotted against the negative logarithm of the dissociation constants of the conjugate acid ($pK_a$) of the nucleophiles. Again, a linear relationship is found, expressed by Equation 3.15:

$$\log k_{nu} = g + \beta pK_a^{(nu)} \quad (3.15)$$

An example is seen in Figure 3.3 (open circles only). In such a plot, the slope represents the sensitivity of the rate to the strength of the acid or the nucleophile, whether it is acting as a catalyst or as a reactant.

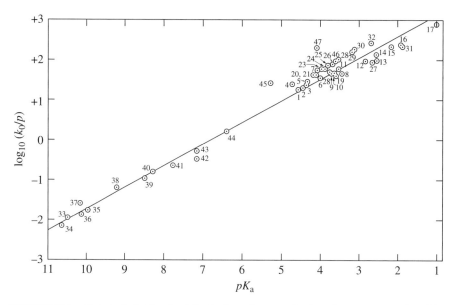

**FIGURE 3.2**   Brønsted plot for the dehydration of acetone hydrate in acetone, catalyzed by acid. The logarithm of the catalytic rate constant $k_a$ is plotted versus the negative of the dissociation constant of the acid. Points are shown for 32 carboxylic acids and 15 phenols. The point labeled 47 represents 2,4-dinitrophenol. Acids of other types (e.g., oximes) fell much further off the line (see Section 3.6.3). Source: Bell and Higginson (1949) by permission of the Royal Society of London.

### 3.6.3   Solvent-Effect Studies on the Alpha Effect

In certain types of reactions involving nucleophiles, it has been found that rate constants for nucleophiles that have an atom bearing a lone pair of electrons adjacent to ("alpha to") the nucleophilic atom, when included in a Brønsted-type plot, lie above the line, that is to say, their nucleophilic strength is enhanced above what is expected for a normal nucleophile of the same base strength. Examples of such "supernucleophiles" are illustrated by the filled circles in Figure 3.3. This enhancement is termed the *alpha effect* (See, e.g., Buncel *et al.*, 1982). It is measured by the ratio $k$(alpha Nu)/$k$(normal Nu). Note that the "normal Nu" is one (which may be hypothetical) with the same $K_a$ as the "alpha Nu." Commonly found alpha nucleophiles contain as the "alpha" atom:

- Oxygen (e.g., $HOO^-$, $CH_3OO^-$, $CH_3C(O) COO^-$, $MeONH_2$)
- Nitrogen ($NH_2-NH_2$, oximes)
- Others ($ClO^-$, $BrO^-$ $HSO_3^-$)

Reactions at $sp^3$ centers are associated with small (<10) alpha effects, as typified by those illustrated in Figure 3.3. On the other hand, the rather small value of the alpha effect for the $S_N2$ reaction of methyl *p*-nitrophenyl sulfate is contrasted with the reaction of *p*-nitrophenyl acetate at an $sp^2$ center, with butane-2,3-dionemonoximate ($Ox^-$)

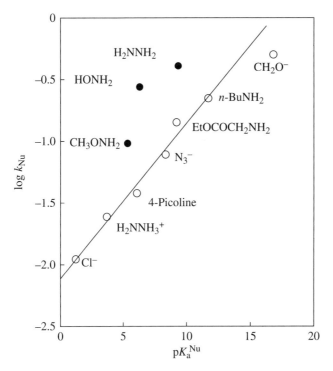

**FIGURE 3.3** Brønsted-type plot of log $k_{Nu}$ versus p$K_a$ for the reaction of nucleophiles with methyl *p*-nitrophenyl sulfate. The $\alpha$-nucleophiles are shown as solid circles. Source: Buncel *et al.* (1980) by permission of the American Chemical Society.

as an alpha nucleophile and 4-ClPhO⁻ as normal Nu, with an alpha effect of 100 (Equation 3.16). Even larger alpha effects are observed for reactions at sp centers.

$$MeC(=O)-O-\!\!\bigcirc\!\!-NO_2 + Nu^- \longrightarrow$$

**PNPA**

(3.16)

$$MeC(=O)-Nu + {}^-O-\!\!\bigcirc\!\!-NO_2$$

$$Nu^- = PhC(CF_3)=NO^-(TFA\text{-}Ox^-) \text{ and } 4\text{-}ClC_6H_4O^- \text{ (4-ClPhO}^-)$$

Intense studies have been directed to tracing the origin of the alpha effect, but so far, this has been elusive. Four possible origins have been generally considered so far: (i) ground-state destabilization, (ii) transition state stabilization, (iii) product stabilization, and (iv) solvent effects. Herein, we consider primarily the solvent-effect factor, while for the more general background, the reader is referred to several reviews, for example, Castro (1999), Johnson (1975), and Parker (1969).

One type of solvent-effect study considers diverse solvents such as water, methanol, acetonitrile, toluene, dimethyl sulfoxide (DMSO), and so on. These studies showed a variation in the alpha effect but, as may have been expected, did not point to a single factor such as solvent polarity as causing the rate enhancement. Interestingly, it was the variation in solvent composition of binary solvent mixtures, such as DMSO–$H_2O$ and MeCN–$H_2O$, that gave the most valuable clues concerning the origin of this solvent effect of alpha nucleophiles.

The manner in which the alpha effect is exhibited can be enlightening. The systematic studies mentioned earlier have revealed different forms of curves in plots of solvent-dependent alpha effects: bell-shaped (Buncel and Um, 1986), increasing (Kwon *et al.*, 1989), and even decreasing (Um and Buncel, 2001; Um *et al.*, 1993) as the proportion of organic solvent in aqueous/organic mixed solvent increases. It is clear that measurements in the pure solvents alone can be misleading, as they only provide two points on what may be a complex curve. Examples of these forms are shown in Figures 3.4 and 3.5. Many studies have involved aqueous DMSO mixtures, but it should be emphasized that the DMSO–$H_2O$ system is not unique in displaying

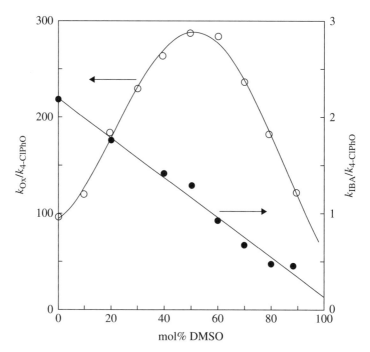

**FIGURE 3.4**   Plots showing the effect of solvent composition on the $\alpha$-effect for the reactions of PNPA with Ox⁻ (an alpha nucleophile) versus 4-ClPhO⁻ (a normal nucleophile), open circles, left-hand scale, and with IBA⁻ versus 4-ClPhO⁻, filled circles, right-hand scale, in DMSO–water mixture at 25.0±0.1°C. Source: Buncel and Um (2004) with permission of Elsevier Ltd.

the composition dependence of alpha effect. An example of the effect in MeCN–H$_2$O mixtures is seen in Figure 3.5 (the rising curve).

A commonly observed form of curve is bell-shaped. Figure 3.6 compares the effect observed in the reaction of a single alpha nucleophile (Ox$^-$) at several common electrophilic centers: C=O (PNPA, 4-nitrophenyl acetate), P=O (PNPDDP, 4-nitrophenyl diphenyl phosphinate), and SO$_2$ (PNPBS, 4-nitrophenyl diphenyl sulfonate). Each of these shows the characteristic bell-shaped plot; however, the magnitude of the alpha effect covers a 10-fold range.

One can dissect the overall solvent effect for a reaction pathway in two different media into the effect of solvent change on the reactants, R, and on the transition state T. Consider an example of the reaction of a normal nucleophile with a substrate, in two solvents, such as H$_2$O (0) and DMSO–H$_2$O (S), as illustrated schematically in Figure 3.7. Equation 3.17, which simplifies to Equation 3.18, affords a link between the measured difference between the activation Gibbs energies, $\delta \Delta G_{tr}^{\ddagger}$; the transfer Gibbs energy of the reactant(s), $\delta G_{tr}^{R}$ (which can be obtained from

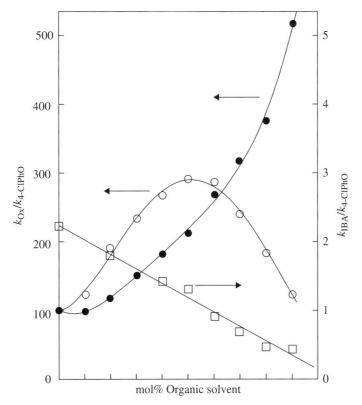

**FIGURE 3.5** Effect of solvent composition on the $\alpha$-effect for reactions of PNPA at 25.0±0.1°C: $k$ (Ox$^-$)/$k$(4-ClPhO$^-$) in MeCN–H$_2$O ($\bullet$); $k$ (Ox$^-$)/$k$(4-ClPhO$^-$) in DMSO–H$_2$O (○); and $k$ (IBA$^-$)/$k$(4-ClPhO$^-$) in DMSO–H$_2$O (□). Source: Buncel and Um (2003) with permission of Elsevier Ltd.

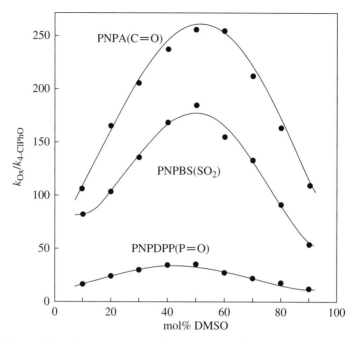

**FIGURE 3.6** Effect of solvent composition on the $\alpha$-effect for the reactions of PNPA, PNPDPP, and PNPBS with Ox$^-$ versus 4-ClPhO$^-$ in DMSO–water mixture at $25.0 \pm 0.1\,^\circ$C. Source: Buncel and Um (2003) with permission of Elsevier Ltd.

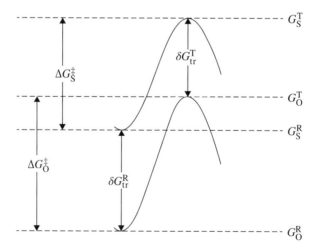

**FIGURE 3.7** Illustration of the relationship between the transfer free energies of reactant and the transition state and the free energies of activation for a reaction occurring in two solvent systems. The reference solvent, in this case water, is denoted by the zero subscript, and the solvent under investigation (such as DMSO–water mixture) by subscript S. Uppercase $\Delta$ refers to change during activation; lowercase $\delta$ refers to change of solvent. Source: Buncel and Um (2003) with permission of Elsevier Ltd.

independent measurements); and the desired transfer Gibbs energy of the transition state, $\delta G_{tr}^T$ (Um *et al.*, 2006):

$$\delta \Delta G_{tr}^{\ddagger} = \Delta G_S^{\ddagger} - \Delta G_O^{\ddagger} = \left( G_S^T - G_S^R \right) - \left( G_O^T - G_O^R \right) \tag{3.17}$$

$$\delta \Delta G_{tr}^{\ddagger} = \delta G_{tr}^T - \delta G_{tr}^R \tag{3.18}$$

Repeating this analysis on the reaction of an alpha nucleophile with the same substrate then enables one to see how the change of solvent affects each of the four quantities that combine to cause the variation of the alpha effect as solvent composition changes, namely, $\delta G_{tr}^T(\text{alpha})$, $\delta G_{tr}^R(\text{alpha})$, $\delta G_{tr}^T(\text{normal})$, and $\delta G_{tr}^R(\text{normal})$. Each may be affected by change in solvent composition differently, and the resulting form of curve may be complex.

### 3.6.4 The Kinetic Isotope Effect

If a reacting species with the natural isotopic composition is changed by substitution of one or more atoms by a less-common (usually heavier) isotope, a *kinetic isotope effect* (KIE) is observed. Simply viewed, the effect may be related to the difference between the zero-point energies (ZPEs) of the vibrational mode principally involving the substituted atom, in the initial and transition states. Assuming harmonic motion, the classical frequency of vibration of a bond between atoms of masses $m_1$ and $m_2$ with force constant k is given by Equation 3.19, in which $\mu$ is the reduced mass, $m_1 m_2/(m_1 + m_2)$:

$$\nu = \left( \frac{1}{2\pi} \right) \sqrt{\frac{k}{\mu}} \tag{3.19}$$

The ZPE is then given by Equation 3.20:

$$E_0 = \frac{1}{2} h\nu \tag{3.20}$$

To a good approximation, the Morse curve of potential energy versus interatomic distance is the same for different isotopic masses, so the force constant will also be essentially the same. If the atom being substituted is attached to a substantially heavier atom, that is, if $m_1 \ll m_2$, then $\mu \approx m_1$. The relative difference in the square roots of the masses of the isotopes is thus important, so the tritium isotope effect is potentially the largest, closely followed by deuterium. Heavy atom KIEs, while smaller in magnitude, are nevertheless readily measurable and can be highly informative of reaction mechanism (Buncel and Saunders, 1992). Reactions in which the rate-determining step involves breaking the bond to the substituted atom are reduced in rate by substitution of a heavier isotope. This is the normal, *primary KIE*.

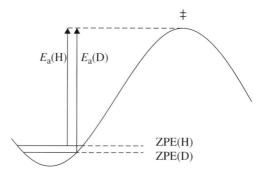

**FIGURE 3.8** Energy profile illustrating the effect of the zero-point energy of the initial state only, on the activation energy for carbon–deuterium versus carbon–hydrogen bond scission, leading to a primary isotope effect. Source: Buncel and Dust (2003) with permission from the American Chemical Society.

Consider a reaction step in which a C–H (or C–D) bond is broken, and H is transferred from a molecule R–H to a base B. The reaction profile for the simple model is represented in Figure 3.8. Only the C–H stretch in the initial state is considered; the transition state (TS) is ignored. The vibrational energy of the C–D stretch lies below that for C–H by the difference between their ZPEs, so $\nu_D/\nu_H \approx 0.71$. Estimating the force constant from typical R–H stretching frequencies around $3000\,cm^{-1}$, we can estimate the rate constant ratio $k_H/k_D \approx 7$ at 25°C, as a maximum value.

Other modes, in either the initial or transition states, can influence the outcome. To consider the TS first, if it is linear (Fig. 3.9), there are two modes involving the C–H and H–B bond distances. The antisymmetric mode (Fig. 3.9a) corresponds to motion along the reaction coordinate and is not associated with a ZPE. In the symmetric stretch (Fig. 3.9b), the proton is nearly stationary (unless the masses of the C and B atoms are very different or the proton is off-center), so this mode may also be neglected. A nonlinear transition state, however, will have a symmetric stretch in which the proton moves, so the difference between the ZPEs for the TSs containing H and D will not be negligible, reducing the KIE. Inclusion of this and other modes whose ZPEs are altered by isotopic substitution in the initial and transition states (Fig. 3.10) can lead to KIEs differing from the value 7 derived from the simple model.

Secondary isotope effects, which are generally smaller, are observed if the substituted atom is not directly involved. Several extensive theoretical treatments of KIEs are available to the reader (Buncel and Dust, 2003; Collins and Bowman, 1970; Melander and Saunders, 1980; see also the series, *Isotopes in Organic Chemistry*, Buncel and Saunders, Eds. 1992, and earlier volumes in the series, Buncel and Lee, Eds.).

The *kinetic solvent isotope effect* (KSIE) is almost exclusively studied in terms of deuterium effects, because it arises naturally in the context of studies of general acid–base catalysis (Alvarez and Schowen, 1987; Kresge *et al.*, 1987). Smith and March (2007, p. 327) as well as Carroll (1998, p. 365) briefly describe the factors that may lead to KSIEs, including direct participation of the solvent as a reactant, exchange of

(a)

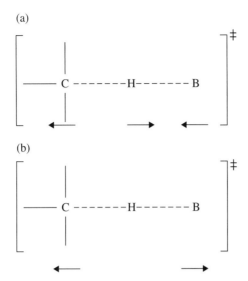

(b)

**FIGURE 3.9** Linear activated complex. (a) Antisymmetric stretch: the H is being transferred from C to B. (b) Symmetric stretch: the H hardly moves. Source: Buncel and Dust (2003) with permission from the American Chemical Society.

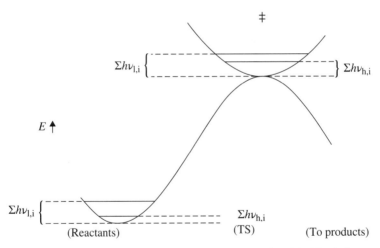

**FIGURE 3.10** A general energy profile that illustrates the origin of the kinetic isotope effect in terms of the zero-point energies in both the initial and transition states. The ZPE for the heavy isotope (subscript h) and for the light isotope (subscript l) arise from summation over the various vibrational modes (i). Source: Buncel and Dust (2003) with permission from the American Chemical Society.

deuterium between solvent and substrate before reaction, and differential solvation of reactants and transition state. The largest KSIE is expected when the rate-limiting step of a reaction depends on breaking an O–H bond of the solvent, but two or all three factors may be simultaneously active. Alvarez and Schowen (1987), Kresge *et al.* (1987), and More O'Ferrall *et al.* (1971) discuss the separation of these factors. See also "Secondary and Solvent isotope effects," vol. 7 in the series, *Isotopes in Organic Chemistry*, Buncel and Lee, Eds. (1987). Stein and Würzberg (1975) describe an effect on the fluorescence of rare-earth ions in $H_2O$ and $D_2O$ that may be described as accidental; if the difference in energy between the highest nonfluorescing level of the ion and its lowest fluorescing level happens match the energy of an excited OH or OD stretch, the fluorescence is strongly quenched. Notably, the fluorescence yield of europium(III) in $H_2O$ and $D_2O$ differs by a factor of 40.

There is currently much activity in the field of KSIEs. A search using this phrase in 2002 yielded 118 references to work on their use in elucidating a large variety of reaction mechanisms, nearly half in the preceding decade, ranging from the $S_N2$ process (Fang *et al.*, 1998) to electron transfer in DNA duplexes (Shafirovitch *et al.*, 2001). Nineteen countries were represented: see, for example, Oh *et al.* (2002), Wood *et al.* (2002), Blagoeva *et al.* (2001), and Koo *et al.* (2001). A similar (unfiltered) search in 2013 yielded over 25,000 "hits."

## 3.7 LEWIS ACIDS AND BASES

Lewis bases are species that can provide a pair of electrons to be shared with another species. Lewis bases are also Brønsted bases, for a proton can be the "other species." A Lewis acid, which is a species that can receive a share in a pair of electrons, cannot act directly as a Brønsted acid, but it can manifest acidity in an amphiprotic solvent by forming an adduct with a solvent molecule that is then a stronger acid than the solvent, for example,

$$Al^{3+} + 7H_2O \rightarrow [Al(OH_2)_6]^{3+} + H_2O \rightleftharpoons [Al(OH_2)_5OH]^{2+} + H_3O^+$$

Thus, very strong Lewis acids cannot exist in appreciable concentration in an amphiprotic solvent. (Nor can very strong Brønsted acids, owing to the leveling effect.)

Various measures of Lewis base strength have been proposed. One such is the *donor number* (Gutmann, 1968, 1969; Gutmann and Wychera, 1966), defined as the negative of the enthalpy of reaction of the base with the strong Lewis acid, antimony pentachloride, in dilute solution in 1,2-dichloroethane. This solvent was chosen as it is nearly (though not quite) nonbasic. In fact, as Gutmann (1968, p. 19) points out, as well as the binding effect of the unshared electron pair, the donor number includes any dipole–dipole or ion–dipole contributions and steric effects. It can be used to rationalize solubilities both of (Lewis) acids in basic solvents and of bases in acidic solvents and differences in reactions of certain substances in different solvents. For instance, Gutmann (1968, p. 148) cites work by Mikhailov and coworkers showing that boron(III) chloride can donate a chloride ion to iron(III) chloride in diethyl ether (DN = 19.2) or in tetrahydrofuran (DN = 20.0) according to the reaction

$$BCl_3 + FeCl_3 + 2(C_2H_5)_2O \rightleftharpoons [Cl_2B(O(C_2H_5)_2)_2] + [FeCl_4]^-$$

but not in phosphorus oxychloride (DN = 11.7) and only to a certain extent in phenylphosphonic dichloride (DN = 18.5). Stabilization of $[BCl_2]^+$ requires a certain degree of donor strength on the part of the coordinating solvent. The concept of leveling, discussed earlier, is also used by Gutmann (1968, p. 168) in considering the effect of the coordinating power of the solvent, as measured by the donor number, toward transition-metal complex formation. In solvents of low donor number, such as nitromethane (DN = 2.7), reactions of the type exemplified by

$$Fe^{3+}(solv) + 4X^- \rightleftharpoons [FeX_4]^-$$

where $X^-$ may be any of $I^-$, $Br^-$, $Cl^-$, $N_3^-$, or $NCS^-$, are all nearly complete when the reactants are present in stoichiometric amounts. Such reactions are thus leveled in this solvent. On the other hand, in DMSO (DN = 29.9), $FeCl_3$ is completely solvo-lyzed to $Fe(dmso)_n^{3+}$ and $Cl^-$, though the tetra-azido and hexathiocyanato complexes of $Fe^{3+}$ are stable. DMSO is differentiating in this sense.

## 3.8  HARD AND SOFT ACIDS AND BASES (HSAB)

Ahrland *et al.* (1958) introduced the *a/b* classification of metal cations. *Class (a)* contains those cations that form their most stable complexes with the ligand atoms of the first member of each group of the periodic table, that is, with F, O, or N rather than Cl, S, or P. *Class (b)* contains those ions that display the reverse order of stability. The reversal of the order of affinity of the halide ions as bases for the two acids $Al^{3+}$ ($I^- \ll F^-$) and $Hg^{2+}$ ($I^- \gg F^-$) is an example. Pearson (1963, 1988), in an attempt to account for these and related observations, noted that the small, highly charged ion $Al^{3+}$ was most tightly bound to the small ion $F^-$, and the larger, less charged ion $Hg^{2+}$ best bound the larger $I^-$, in keeping with Fajans's (1923) rules. He described as *hard*, acids in which the vacant orbital that accepts an electron pair is (when occupied) low-lying and relatively unpolarizable, and bases in which the donatable pair is similarly in a small, low-lying orbital and relatively unpolarizable. *Soft* acids and bases have relevant orbitals that are not so low in energy and are relatively large and polarizable. Overlap is greatest and the interaction strongest when the orbitals involved are most closely similar in size and energy. Thus, *hard acids react most strongly with hard bases, and soft with soft*. The presence of soft ligands tends to induce softness in a central atom: $BF_3$ is hard, but $B(CH_3)_3$ is borderline. In ligand reorganization reactions of the type

$$MA_n + MB_n \rightleftharpoons MA_jB_k + MA_kB_j$$

(where $j + k = n$) if A and B differ in softness, the ligands might be expected to show discrimination[3] as defined in Chapter 2. Such cases have been reported, for example, the system $Si[N(CH_3)_2]_4$ and $SiCl_4$ (Basolo and Pearson 1967, p. 445). Clark and Brimm (1965), however, found the distribution of the ligands $PF_3$ and CO among the five possible compounds resulting from the equilibration of a mixture of $Ni(PF_3)_4$ and $Ni(CO)_4$ to be essentially random, though $PF_3$ and CO are by no means equally soft. The stability of methylmercuric chloride toward disproportionation into $HgCl_2$ and $Hg(CH_3)_2$ (Buncel et al., 1986), an example of *antidiscrimination*, is explained by invoking subtleties of bond hybridization, implying competition between d orbitals on two chlorines for the same mercury orbital. *Discrimination* is favored by difference in softness of the ligands and disfavored by great difference in their base strength (Pearson, 1973, pp. 51, 77). Considering, hypothetically, the halogenated methanes as formed from $C^{4+}$ (hard), $H^-$ (soft), $F^-$ (hard), and $Cl^-$ (intermediate), the hard and soft acids and bases (HSAB) theory would predict a greater degree of discrimination in the $CH_nF_{4-n}$ system than in the $CH_nCl_{4-n}$ system. The reactions are of course immeasurably slow, but the equilibrium constant calculated from Gibbs energies of formation for the reaction

$$2CH_2X_2 \rightleftharpoons CH_4 + CX_4$$

with $X=F$ is $8.5 \times 10^{15}$ and with $X=Cl$ is $6.1 \times 10^{-4}$. The constant corresponding to unperturbed statistics is 0.028, so substantial discrimination exists in the fluoromethanes. The chloromethanes show a degree of antidiscrimination, perhaps merely due to crowding of the chlorines in $CCl_4$. The absence of d orbitals of low energy on carbon makes the argument using chlorine d orbitals less tenable.

Gritzner (1997) has recalled Ahrland's caution that a soft–soft interaction is not as simple as the hard–hard interaction as described by Lewis, that is, donation of share of an electron pair of the base to a vacant orbital on the acid to form a σ bond, but it may also involve π back-donation from the acid to the base. There is a difference in kind as well as strength of interaction. Nevertheless, many authors have found it convenient to consider all Lewis-type acid–base reactions as lying on a spectrum from fully hard, exemplified by hydrogen bond formation, to very soft, typified by mercury–sulfur bonding, and have endeavored to construct scales of hardness or softness. They are considered in the next section.

## 3.9  SCALES OF HARDNESS OR SOFTNESS

Edwards (1954, 1956), in a study of Lewis acid–base reactions involving metal cations and anionic ligands, showed that the equilibrium constants can be correlated by Equation 3.21:

$$\log\left(\frac{K}{K_0}\right) = \alpha E_n + \beta H \tag{3.21}$$

in which the metal ions are characterized by a pair of parameters $\alpha$ and $\beta$ and the ligands by $E_n$ and $H$.

The ligand parameters are derived from independent properties of the ions $Y^-$: $E_n$ from the standard oxidation potential for the reaction

$$Y^- = \tfrac{1}{2}Y_2 + e^-$$

and $H$ from p$K$ for the ionization of HY.

Figure 3.11 shows a plot of the values of $\beta$ versus $\alpha$ for 17 metal ions (Yingst and McDaniel, 1967; cited by Jolly 1970, p. 53). Superimposed is a suggestion of part of a new coordinate system (effectively a rotation of the original coordinate system), *softness* $S_+^A$ (and *strength*, not shown). Except for the point for $Zn^{2+}$, which is classified by Pearson as "borderline" but which nearly coincides with the hard $Mg^{2+}$, the *zero softness* line (the "borderline") divides the hard from the soft ions. The softness coordinate shown is approximately represented by Equation 3.22. Values of $\alpha$, $\beta$, and $S_0^A$ are listed in Table 3.3:

$$S_+^A = 0.465\alpha - 1.838\beta - 0.623 \tag{3.22}$$

Reactions of the type $A + :B \to AB$ where A and :B are a Lewis acid and base have been studied by calorimetry (in a poorly coordinating solvent) by Drago and Wayland (1965). They showed that the enthalpy of reaction between uncharged species can be expressed by Equation 3.23:

$$-\Delta H = E_A E_B + C_A C_B \tag{3.23}$$

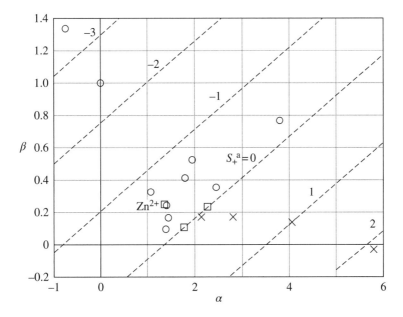

**FIGURE 3.11** Edwards's parameters, $\beta$ versus $\alpha$, for 17 cations, with a proposed softness scale superposed. Circles represent hard ions, crosses soft, and squares borderline. Note the apparently anomalous position of $Zn^{2+}$. Data from Yingst and McDaniel (1967).

**TABLE 3.3   Comparison of $\alpha/\beta$ with Softness. Edwards's Acid Parameters for 17 Cations and the Tentative Values of the Proposed Softness Parameter $S_+^A$ (see Text)[a]**

| Ion | $\alpha$ | $\beta$ | $S_+^A$ | H/S | Ion | $\alpha$ | $\beta$ | $S_+^A$ | H/S |
|---|---|---|---|---|---|---|---|---|---|
| $Hg^{2+}$ | 5.786 | −0.031 | 2.1 | S | $Ga^{3+}$ | 3.795 | 0.767 | −0.3 | H |
| $Cu^+$ | 4.060 | 0.143 | 1.0 | S | $Mg^{2+}$ | 1.402 | 0.243 | −0.4 | H |
| $Ag^+$ | 2.812 | 0.171 | 0.4 | S | $Zn^{2+}$ | 1.367 | 0.252 | −0.5 | B[b] |
| $Cd^{2+}$ | 2.132 | 0.171 | 0.1 | S | $Ba^{2+}$ | 1.786 | 0.411 | −0.5 | H[c] |
| $Pb^{2+}$ | 1.771 | 0.110 | 0.0 | B | $Fe^{3+}$ | 1.939 | 0.523 | −0.7 | H |
| $Cu^{2+}$ | 2.259 | 0.233 | 0.0 | B | $Ca^{2+}$ | 1.073 | 0.327 | −0.7 | H |
| $In^{3+}$ | 2.442 | 0.353 | −0.1 | H | $H^+$ | 0.000 | 1.000 | −2.5 | H |
| $Sr^{2+}$ | 1.382 | 0.094 | −0.2 | H | $Al^{3+}$ | −0.749 | 1.339 | −3.4 | H |
| $Mn^{2+}$ | 1.438 | 0.166 | −0.3 | H[c] | | | | | |

[a] Data from Yingst and McDaniel (1967).
[b] Anomalous.
[c] Not classified by Pearson.

Acid and base strengths are represented not by a single number for each, but rather two numbers, $E_A$ and $C_A$, to characterize the acid, and similarly $E_B$ and $C_B$ to characterize the base. They interpret the $E$'s as measuring the tendency to electrostatic bonding and the $C$'s as measuring the tendency to covalent bonding. Some values of the parameters are tabulated in the Appendix. As with the Pearson and Edwards descriptions, discussed earlier, the Pearson and Drago descriptions are also related somewhat like two coordinate systems in a plane, differing in orientation, and with some displacement of the origin (see Fig. 3.12).

It is probably too much to expect that single scale of acid or base softness (or hardness) could be established for all acid or all basic species. Drago *et al.* (1991), in considering reactions in which one or both reacting species are charged, find that it is necessary to introduce an additional term $R_A T_B$ to Equation 3.23, to account for electron density transferred. This complicates the interpretation of softness as one of a pair of rotated coordinates. Where both species are uncharged, the $R_A T_B$ term is small, so only uncharged species were considered in the construction of Figure 3.10. It may be possible to define a set of softness scales for acids, $S_v^A$, where v is the charge on the species or at the active site, and a similar set, $S_v^B$, for bases. Huheey *et al.* (1993, p. 349) present a table of values of equilibrium constants (as p$K$) for a variety of neutral and anionic bases reacting with the hard acid $H_3O^+$ and the soft acid $CH_3Hg^+$. There are clear cases where the strength of the base can be related to the sum of the p$K$'s and the softness to their difference, but the presence in the list of bases of univalent and divalent anions, of simple neutrals like ammonia, with a single basic site, and species like $SCN^-$, DMSO, and $Et_2PC_2H_4OH$, which have two sites of widely differing softness, make the interpretation difficult.

Marcus (1987) defines the (base) softness parameter $\mu$ for a solvent as "the difference between the mean of the Gibbs free energies of transfer of sodium and potassium ions from water to a given solvent and the corresponding quantity for silver ion (/kJ mol$^{-1}$), divided by 100." He points out that $\mu$ is uncorrelated with the $\beta$ scale

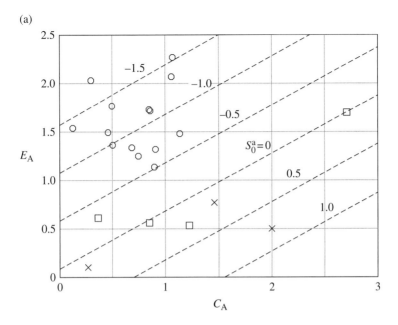

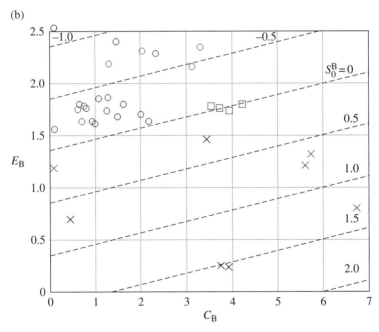

**FIGURE 3.12** (a) Drago–Wayland parameters $E_A$ and $C_A$ for nonionic acids. Hard acids are shown as circles, soft as crosses, and borderline as squares. The bonding atom is indicated by color: black for hydrogen and red for others. Lines are superposed representing values of a proposed softness parameter, defined by $S_0^a = 0.088 - E_A + 0.595 C_A$. (b) Drago–Wayland parameters for nonionic bases: $E_B$ versus $C_B$. Colors indicate bonding atoms: red O, green N, and black others (S, P, C, Cl). Hard bases are shown as circles, soft as crosses, and borderline as squares. The dashed lines are for indicated values of the proposed softness parameter for neutral bases, defined by $S_0^B = 1.35 - E_B + 0.109 C_B$.

**TABLE 3.4 Trilinear Correlation of the Marcus Softness Parameter $\mu$ with Either the Volume Refraction $R_v$ or Molar Refraction $[R]^a$ and with Swain's Parameters $A_j$ and $B_j$**

|            | $C_0$          | $C_1$             | $C_2$         | $C_3$          |
|------------|----------------|-------------------|---------------|----------------|
| Using $R_v$ | $-0.4 \pm 0.4$ | $2.3 \pm 1.4$     | $-0.07 \pm 0.19$ | $-0.04 \pm 0.22$ |
| Using $[R]$ | $-0.08 \pm 0.10$ | $0.0094 \pm 0.0045$ | $0.05 \pm 0.18$ | $-0.01 \pm 0.22$ |

$^a R_v = (n^2 - 1)/(n^2 + 2); [R] = R_v M/\rho.$

of solvent basicity (Kamlet and Taft, 1976), which is a measure of hydrogen-bond acceptor strength, "hard" basicity. It might be expected that $\mu$ might show some correlation with the volume refraction, $R_v = (n^2 - 1)/(n^2 + 2)$, since softness is considered to be related to polarizability. Multilinear correlation of $\mu$ with $R_v$ and Swain's *acity* and *basity* parameters, $A_j$ and $B_j$, in the form

$$\mu = C_0 + C_1 R_v + C_2 A_j + C_3 B_j$$

yielded values of constants listed in Table 3.4. Softness is not a bulk property, but a molecular one, or more precisely a property associated with the acidic or basic site within a molecule. A close correlation with $R_v$ (a bulk property) or with $[R]$, the molar refraction (Lorentz–Lorenz) should not be expected. Nevertheless, both were tried. Clearly, the correlation coefficients with $A_j$ and $B_j$ are zero in either form. The correlation with $[R]$ is only slightly better (relative standard deviation of slope=0.60) than with $R_v$ (0.41).

Four solvents (ammonia, acetonitrile, pyridine, and triethylamine) did not fit either correlation, $\mu$ for these lying some 0.2–0.5 units above the trend of the remainder. That they are all nitrogen compounds that have a particular affinity for silver ion may be significant. In a similar correlation of $\mu$ with Drago's $C_b$ and $E_b$ for the very few solvents for which all three quantities are known, ammonia lay 0.6 units above the trend. (Two other nitrogen compounds, $N,N$-dimethylformamide and hexamethylphosphoramide, are not apparently anomalous.) Softness is also expected to be related to the energy gap between the *highest occupied* and *lowest unoccupied molecular orbitals* (HOMO and LUMO). These are also molecular properties, so their relevance to the local properties at acid or base sites varies with the complexity of the molecule. For atoms, or for small molecules with a central atom and simple ligands, softness should be strongly correlated with the reciprocal of the HOMO–LUMO gap. The comparison of several base softness scales proposed by Chen et al. (2000), and others, is discussed in Chapter 4.

## 3.10   ACIDS AND BASES IN REACTIVE APROTIC SOLVENTS

The term *autoionization* (but not *autoprotolysis*) may be applied to reactions of certain aprotic solvents, for example,

$$2SbCl_5 \rightleftharpoons SbCl_4^+ + SbCl_6^-.$$

Here, the ion transferred is Cl⁻, rather than H⁺. Addition of a strong Lewis acid to such a solvent will result in the appearance in solution of the characteristic positive ion of the solvent. Addition of a base will yield the negative ion. Again, aprotic solvents may be leveling toward acids stronger than the characteristic acid ion of the solvent or, mutatis mutandis, toward bases. It will be seen that this is consistent with the picture based on the proton.

Generally, bases are recognized by their ability to provide negative charge in the form of actual negative ions or of electron pairs to be shared. Acids are recognized by their ability to provide positive charge or to accept shared electron pairs. In the context of reaction kinetics and mechanisms, a base is termed a *nucleophile*, and a Lewis acid an *electrophile*.

### 3.11 EXTREMES OF ACIDITY AND BASICITY

J. B. Conant (Hall and Conant, 1927) recognized that certain systems, such as Friedel–Crafts catalysts, can have acidities greater than that of 100% $H_2SO_4$ ($H_0 = -12$). These Gillespie and Peel (1971, 1973) called *superacids*. The strongest known (Gillespie and Peel, 1973) simple Brønsted acid is fluorosulfonic acid, $FSO_3H$, with $H_0 = -15.1$ for the neat acid, closely followed by trifluoromethanesulfonic acid ("triflic acid") at −14.1. Even greater acidities can be achieved by the formation of complexes between certain Brønsted and Lewis acids. A solution of $1\,M\,SbF_5$ in HF has been shown to have a value of $H_0$ beyond −22 (Fabre *et al.*, 1982; Olah *et al.*, 1985). These acid systems extend greatly the possibilities for acid catalysis and make possible the preparation of solutions containing carbocations, stable enough for study.

The complementary term is *superbase*. Solvents that are sufficiently basic (e.g., ammonia) or at least weak enough acids (e.g., DMSO) can support high levels of basicity. Jolly (1970, p. 104) shows a figure representing the effective pH ranges (water scale) possible in several solvents, extending to 30 in DMSO and to 37 in ammonia. Amide ion in ammonia can deprotonate molecular hydrogen enough to catalyze the exchange reaction

$$HD + NH_3 \rightleftharpoons H_2 + NH_2D$$

which has been considered as a basis for deuterium enrichment (Buncel and Symons, 1986).

### 3.12 OXIDATION AND REDUCTION

The strength of an oxidant or a reductant that can exist for an indefinite period in any given solvent is obviously limited by the susceptibility of the solvent to oxidation or reduction. One must, however, consider both the thermodynamic and the kinetic aspects of any possible reaction. For instance, in aqueous solution in the absence of catalysts, oxidants that should, thermodynamically, be able to drive the half-reaction

$$2H_2O \rightleftharpoons 4H^+ + O_2 + 4e^-$$

can nevertheless exist in solution as useful reagents: permanganate, for instance. Similarly, the reaction

$$2Cr^{2+}(aq) + 2H^+(aq) \rightleftharpoons 2Cr^{3+}(aq) + H_2(g); \quad K = 9 \times 10^6$$

is so slow that an aqueous solution containing chromous ion may be used as a very strong reductant, as long as oxygen is excluded.

Liquid ammonia dissolves the alkali metals without immediate reaction and affords a strong reducing agent, as long as catalysts (chiefly finely divided metals) for the reaction

$$2NH_3 + 2e^-(am) \rightarrow 2NH_2^- + H_2$$

are excluded.

Barthel and Gores (1994, p. 20) list 23 solvents, selected from several classes, that can sustain large ranges of electrochemical potential, depending on the added electrolyte. The largest range reported is 6.8 V (from −3.3 to +3.5 V on Pt vs. Ag/Ag⁺) for ethylene carbonate containing potassium hexafluorophosphate. Coetzee and Deshmukh (1990) warn that impurities, some difficult to remove, may seriously limit the ranges of oxidation/reduction potential, as well as of acidity or basicity.

### 3.13  ACIDITY/REDOX DIAGRAMS

Pourbaix (1949; Pourbaix *et al.*, 1963) diagrams are widely used, especially in geochemistry, to show the conditions of pH and reduction potential in aqueous media in which various species of an element are stable. Figure 3.13 shows such a diagram for Fe species in water at 25°C. The ordinate is the reduction potential with respect to the standard hydrogen electrode. The solid lines correspond to equilibria between two species at equal concentrations or activities; the fields are labeled with the species that are stable within them.

Consider a half-reaction, $\mathbf{O} + m\mathrm{H}^+ + n\mathrm{e}^- \rightleftharpoons \mathbf{R}$, where $\mathbf{O}$ and $\mathbf{R}$ are the oxidized and reduced principal species. From the form of the Nernst equation, Equation 4.7, in which $F$ is the Faraday constant,

$$E = E_0 - \frac{RT}{2.303nF} \log \left( \frac{[\mathbf{R}]}{[\mathbf{O}][\mathrm{H}^+]^m} \right)$$

$$= E_0 - \frac{RT}{2.303F} \frac{m}{n} \mathrm{pH} - \frac{RT}{2.303F} \log \left( \frac{[\mathbf{R}]}{[\mathbf{O}]} \right)$$

it may be seen that the theoretical slope of the line along which the activities of $\mathbf{O}$ and $\mathbf{R}$ are equal is given by $-m/n$, the ratio of the numbers of protons to electrons transferred, times the factor $RT/2.303F$ (though some authors use, as ordinate, $p\mathrm{e} \equiv (2.303F/RT)\,E$, so the slope is just $-m/n$).

The concept can be extended to organic systems. Figure 3.14 shows a similar diagram for isobutane (symbolized as $i\mathrm{C}_4\mathrm{H}$) and derivatives in liquid HF containing

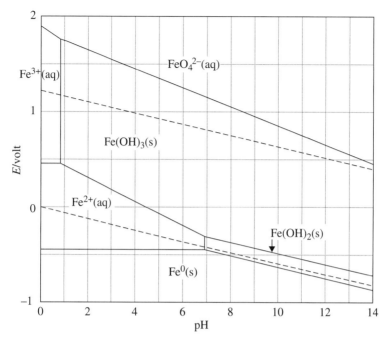

**FIGURE 3.13** Pourbaix diagram for iron species at various reduction potentials $E$ and pH in aqueous environment at 25°C. Only the species indicated in the fields were considered. The dashed lines represent the thermodynamic thresholds for the emission of $O_2$ (upper) and $H_2$ (lower).

$SbF_5$ or KF (Fabre *et al.*, 1982). Here, the ordinate is the potential with respect to the $Ag/AgSbF_6$ electrode, and the abscissa is pH(HF), that is, the pH on a scale with zero corresponding to a 1 M solution of strong acid in HF. The accessible range of pH is that spanned by 0.1 M strong acid (1) to 0.1 M strong base (12.7) in HF, corresponding approximately to $H_0$ from −22 to −8. In the figure, "$iC_4$" represents the *t*-butyl group $(CH_3)_3C$; "$iC_{4=}$" represents isobutene.

In using any of these diagrams, one must be aware that they are constructed on the assumption that only specified species may be present and only specified equilibria are considered. (In Fig. 3.13, for instance, the species $Fe(OH)^{2+}$, $Fe(OH)_2^+$, and $Fe(OH)_4^-$ are omitted, and $H_2FeO_4$ is assumed to be a strong acid.) With that caution, such a diagram may be used to predict the conditions of acidity and reduction potential in which a species is stable with respect to any of the reactions that were considered in the construction of the diagram. A species may be unreactive enough to remain present outside the predicted area of stability. (Most organic compounds are unstable to oxidation by atmospheric oxygen, though the "auto-oxidation" is usually immeasurably slow.)

Trémillon (1971) describes the use of $E$ versus $p(O^{2-})$ diagrams, comparable to $E$ versus pH diagrams, in the study of solutions in fused salts or fused alkali hydroxides. Similar considerations apply.

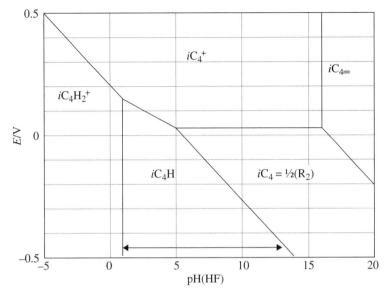

**FIGURE 3.14** Pourbaix diagram for species derived from isobutane (2-methylpropane), represented as $iC_4H$, in solution in HF. The neutral point in HF is at pH=6.9, at which $H_0 \approx -15$. Addition of KF or $SbF_5$ serves to adjust the pH to more basic or acidic values, within the practical limits, 1–13, approximately a range of $H_0$ about −21 to −9 (double arrow). Source: Fabre *et al.* (1982) with permission of the American Chemical Society.

## 3.14   UNIFICATION OF ACID–BASE AND REDOX CONCEPTS

It has been noted by Usanovich (1939) that properties of bases and acids are in a sense parallel to those of reducing and oxidizing agents. In clear cases, it is possible to show that one can formulate an oxidation/reduction reaction as a total transfer of one or more electrons (commonly with such reactions in aqueous solution) or of an atom, and an acid–base reaction as a transfer of an ion. In some circumstances, the distinction may be difficult to define. The acquisition of a fourth oxygen atom by sulfite looks very much like a Lewis neutralization:

$$:\ddot{O}: + :SO_3^{2-} \rightarrow \ :\ddot{O}:SO_3^{2-}$$

with the oxygen atom acting as acid and sulfite as base. The solvated electron, present in the solution of sodium metal in liquid ammonia, can act both as a reducing agent and as a very strong base.

Usanovich took the step of combining the two concepts, and defining acids and bases in terms of all kinds of charge generation or transfer, calling "acids" all species that donate positive charge or accept negative charge, whether by outright transfer or by sharing, and calling "bases" all species that donate negative charge or accept positive charge. Few chemists are willing to follow him so far, as the distinction between acids and bases on the one hand and oxidizing and reducing agents on the other is important. Nevertheless, it is helpful in some cases to keep the parallel nature of these two concepts in mind.

## PROBLEMS

**3.1** The following reactions, carried out in water, have equilibrium constants greater than unity. Identify the acid species involved and sort as many of them as possible into descending order of strength:

1. $NH_3 + H_2CO_3 \rightleftharpoons NH_4^+ + HCO_3^-$
2. $H_2CO_3 + CN^- \rightleftharpoons HCO_3^- + HCN$
3. $H_3PO_4 + CH_3COO^- \rightleftharpoons H_2PO_4^- + CH_3COOH$
4. $HSO_4^- + H_2PO_4^- \rightleftharpoons SO_4^{2-} + H_3PO_4$
5. $H_3O^+ + SO_4^{2-} \rightleftharpoons H_2O + HSO_4^-$
6. $CH_3COOH + NH_3 \rightleftharpoons CH_3COO^- + NH_4^+$
7. $HCN + CO_3^{2-} \rightleftharpoons CN^- + HCO_3^-$

**3.2** When we write for ethanoic acid in water at 25°C, $pK_a = 4.77$, what scale are we implicitly using for the activity of water?

**3.3** For the weak acid, hydrogen phthalate ion $HA^-$, in aqueous solution, the dissociation constant may be written as

$$K = \frac{a_{H^+} a_{A^{2-}}}{a_{AH^-}} = \frac{\left[H^+\right]\left[A^{2-}\right]}{\left[AH^-\right]} \cdot \frac{f_{H^+} f_{A^{2-}}}{f_{AH^-}} = K_c \cdot \frac{f_{H^+} f_{A^{2-}}}{f_{AH^-}}$$

so the "extended" Debye–Hückel equation (with a nod to Davies) for moderate ionic strengths suggests

$$\ln K = \ln K_c + \text{constant}\left(\frac{I^{1/2}}{1 + I^{1/2}}\right)$$

Determine (graphically or otherwise) the best value of $K$ from the following data.

| $I$/mol l$^{-1}$ | 0.0519 | 0.1035 | 0.1551 | 0.2066 | 0.323 | 0.5163 |
|---|---|---|---|---|---|---|
| $K_c \times 10^6$ | 9.11 | 11.55 | 13.72 | 15.73 | 19.52 | 24.81 |

**3.4** In (separate) aqueous sulfuric acid solutions of two Hammett bases, A and B, containing $1.00\,\mu$mol l$^{-1}$ of the respective bases, the tabulated absorbances, due to the base or its conjugate acid, were observed. All readings were taken using a 1.00 cm cell. The value of $pK_a$ for $HA^+$ was $-0.053$. Obtain an estimate of $pK_a$ for $HB^+$ and of the excess acidity $X$ in 1.3, 4.5, and 7.2 M H$_2$SO$_4$. Assume spectroscopic medium effects are absent.

| $C$(acid)/ mol l$^{-1}$ | $C_A = 1\,\mu$mol l$^{-1}$ | | $C_B = 1\,\mu$mol l$^{-1}$ | |
|---|---|---|---|---|
| | $A(\lambda_1)$ | $A(\lambda_2)$ | $A(\lambda_3)$ | $A(\lambda_4)$ |
| 0.01 | 0.773 | 0.034 | 1.005 | 0.009 |
| 10 | 0.002 | 0.506 | 0.011 | 0.705 |
| 1.3 | 0.243 | 0.358 | 0.905 | 0.079 |
| 4.5 | 0.011 | 0.5 | 0.202 | 0.571 |
| 7.2 | 0.003 | 0.506 | 0.023 | 0.696 |

**3.5**  Azoxybenzene (**3**) is stable for indefinite periods in aqueous, alkaline, or dilute acid media. In moderately concentrated sulfuric acid solution, however, it undergoes rearrangement to *p*-hydroxyazobenzene (**4**), a reaction known as the Wallach rearrangement, Equation A:

$$
\begin{array}{c}
\text{O}^- \\
| \\
\text{Ph--}^+\text{N}=\text{N--Ph} \rightarrow \text{Ph--N}=\text{N--PhOH} \\
\quad\; \textbf{3} \qquad\qquad\quad \textbf{4}
\end{array}
\qquad\qquad\qquad\qquad \text{A}
$$

$$
\begin{array}{c}
\text{O}^- \qquad\qquad\qquad\quad \text{OH} \\
| \qquad\qquad\qquad\qquad | \\
\text{Ph--}^+\text{N}=\text{N--Ph} + \text{H}^+ \rightleftharpoons \text{Ph--}^+\text{N}=\text{N--Ph}
\end{array}
\qquad\qquad\qquad \text{B}
$$

A kinetic study of this reaction in media of different acid concentrations (Buncel, 2000) gave the results shown in Table 3.5, which records as well the corresponding $X$ and $H_0$ values. Also given in Table 3.5 are the extents of protonation of azoxybenzene, according to Equation B, which are calculated (Buncel, 1975a, b) from the spectrophotometrically determined value, $pK_a = -5.15$, corresponding to 50% protonation of azoxybenzene in 65% $H_2SO_4$ ($H_0 = -5.15$). From these results, what can you deduce about the role of protonation of **1** in its ease of rearrangement and hence on the mechanism of Reaction A?

**TABLE 3.5   Kinetic Data for the Rearrangement of Azoxybenzene To 4-hydroxyazobenzene At 25°C**

| $H_2SO_4$ (wt%) | $X^a$ | $-H_0^b$ | $\alpha^c$ | $10^5\ k\psi/s^{-1d}$ |
|---|---|---|---|---|
| 75.30 | 5.30 | 6.65 | 0.967 | 0.106 |
| 80.15 | 6.17 | 7.42 | 0.994 | 0.208 |
| 85.61 | 7.20 | 8.35 | 0.999 | 2.17 |
| 90.37 | 8.05 | 9.05 | 1.000 | 7.23 |
| 95.12 | 9.04 | 9.82 | 1.000 | 20.9 |
| 97.78 | 10.01 | 10.35 | 1.000 | 43.8 |
| 99.00 | 10.75 | 10.82 | 1.000 | 76.8 |
| 99.59 | 11.21 | 11.18 | 1.000 | 227 |
| 99.90 | 11.54 | 11.48 | 1.000 | 2310 |
| 99.99 | 11.56 | 11.90 | 1.000 | 4160 |

[a] Data from Buncel (2000).
[b] Cox (1987).
[c] Fraction of azoxybenzene protonated, calculated using $pK_a = -5.15$ (Buncel and Lawton, 1965).
[d] Pseudo-first-order rate constants as determined spectrophotometrically.

# 4

# CHEMOMETRICS: SOLVENT EFFECTS AND STATISTICS

## 4.1 LINEAR FREE ENERGY RELATIONSHIPS

In the previous chapters we have considered the effects of physical properties (cohesive forces, polarity, and polarizability) and chemical properties (chiefly acidity and basicity in their various manifestations) on equilibria and rates of reaction. The problem with theoretical approaches to such matters is that these properties never act alone, nor are they independent. For instance, in a series of solvents that are weakly acidic, the acidity, hydrogen-bonding ability, polarity, and solubility parameter may all be expected to increase more or less together.

Many attempts have been made to bypass the need to disentangle the parallel effects of changing solvents on one phenomenon with another believed to be susceptible to similar influences. The phenomena studied may be chemical equilibria or reaction rates, but they may also be features of electromagnetic spectra: ultraviolet/visible or infrared absorption or nuclear magnetic resonance. By setting up a scale based on one phenomenon, usually a spectral frequency or the logarithm of an equilibrium or rate constant, one may then seek other cases where the data in suitable form, plotted against this scale as abscissa, yield a straight line. The slope of the line may then be interpreted as a measure of the susceptibility of the molecule or the reaction to changes in the particular solvent property on which the scale is based.

This approach is often termed *linear free-energy relationship* (LFER) (Chapman and Shorter, 1972; Hammett, 1970), because it seems to work best when the scales are based on the logarithms of equilibrium constants or rate constants, which are related to the Gibbs free energy of reaction (Eq. 1.1) or of activation (Eq. 1.18), or to changes in spectral frequencies, which are proportional to energy changes. The correlation of substituent effects on rates and equilibria, by Hammett (1937, 1970),

*Solvent Effects in Chemistry*, Second Edition. Erwin Buncel and Robert A. Stairs.
© 2016 John Wiley & Sons, Inc. Published 2016 by John Wiley & Sons, Inc.

is a familiar example. In the present context, the solvent properties that should be considered important are those that have been discussed in the foregoing chapters, namely, general properties such as polarity, the solubility parameter and polarizability, and more specific properties such as hydrogen-bonding ability (whether as donor or acceptor) and Brønsted or Lewis acidity or basicity. One is then concerned with the correlation of these properties with solvent effects on other types of equilibria or on reaction rates. The term *chemometrics* (Wold and Sjöström, 1978) is used to refer to all the various methods of treating chemical data through statistics: correlations of various kinds, and especially *principal component analysis* (PCA), discussed in Section 4.2.

Looking over the array of empirical parameters that have been derived by various authors (see references in Table A.2 and in Reichardt and Welton, 2011, Chapter 7) to correlate effects on reaction rates, equilibria, or spectral frequencies, it appears that there are many effects of the solvent to be considered. The parameters can be divided into two broad categories. First are those that have no "sign," that is, they are in principle symmetric in their effect on cationic or anionic species or on molecules that have electron donor or acceptor properties. These are parameters such as cohesive energy density or cohesive pressure (and its square root, the solubility parameter), internal pressure,[1] polarity, polarizability, refractive index, dielectric constant (relative permeability), and a number of empirical parameters based on particular equilibria, rates, or spectral features. An assortment of these parameters is listed in Table A.2a, with an indication of the experimental basis of each.

The second class contains *dual parameters*, which occur in pairs of complementary attributes: cationic and anionic charge, Lewis or Brønsted acidity and basicity (and refinements such as hard or soft acidity and basicity), electrophilicity and nucleophilicity, and hydrogen-bonding tendency as donor and as acceptor (Table A.2b). A number of the entries in the table are incomplete in that only one of a potential pair of complementary parameters has been investigated. A table of values of most of the listed parameters for selected solvents forms Table A.3.

A further distinction that should be kept in mind is between those parameters that clearly pertain to solvents, such as those derived from bulk properties (refractive index, electrical permittivity, rate and equilibrium constants of reactions in the solvent, UV/visible or IR frequency shifts of solutes), and those that relate to molecules in solution, such as those derived from enthalpies of reactions in dilute solution in an "inert" solvent. The donor number DN and Drago's $E_A$, $E_B$, $C_A$, $C_B$ acidity and basicity parameters for Lewis acids and bases are examples of the latter kind.

Many of the listed parameters are used in equations of the form of Equation 4.1:

$$Z = Z_0 + aA + bB + \ldots \tag{4.1}$$

---

[1]The internal pressure is defined by $P_i = (\partial U / \partial V)_T$. It is of doubtful significance for our purposes; for water below 4°C, for instance, it is negative.

where $Z$ represents the value of a property (usually the logarithm of a reaction rate or equilibrium constant) for a system of interest, $Z_0$ the value of $Z$ for a reference system, the other capital letters represent parameters characteristic of the solvent, and lower-case letters represent coefficients of the corresponding parameters, characteristic of the particular system. An example is the Taft–Kamlet correlation: see Section 4.4.

As noted earlier, the solvent parameters are not independent. Clearly, if a reaction gives rise as a product or as an activated complex to a species that is cationic or has a site with localized positive charge, the reaction will be favored by solvent properties including polarity, polarizability, basicity (whether hard or soft), and by tendencies to covalent or electrostatic interaction with vacant orbitals (i.e., nucleophilicity). Similarly, if the product or activated complex bears a localized negative charge, the reaction will still be favored by solvent polarity and polarizability, but also by acidity and by the presence in solvent molecules of vacant orbitals capable of receiving electron donation (electrophilicity).

Some other relationships are not so obvious; for instance, Dunn *et al.* (1984) found the boiling point and the molar refractivity of some halogenated hydrocarbons to be strongly correlated ($r = 0.92$). Hildebrand's $\delta$ and the polarizability are also correlated. In the simple case of nonpolar, spherical molecules, the London theory of intermolecular forces relates the attractive force between two molecules to the product of their polarizabilities. These forces can also be related to the ionization potentials of the molecules, which are in turn closely related to the energies of the highest occupied molecular orbitals (HOMO). One would hope that by finding out to what extent all these parameters are interrelated, the number of distinct, independent variables could be much reduced. Many of the supposedly symmetric parameters in Table A.2a are related to some of the dual parameters in Table A.2b. The quantity $E_T(30)$, for instance, usually treated as one of the symmetric parameters, contains a strong component of acidity (hydrogen-bond donation).

## 4.2 CORRELATIONS BETWEEN EMPIRICAL PARAMETERS AND OTHER MEASURABLE SOLVENT PROPERTIES

One would like to be able to understand the effects of solvents upon reactions and spectra in terms of properties of molecules or bulk solvents such as the following:

- polarity
- polarizability
- suitable functions of dielectric constant
- corresponding functions of refractive index
- dipole moment
- quadrupole moment
- Lewis acidity and basicity
- hardness or softness
- hydrogen-bond donor or acceptor strength

**TABLE 4.1  Parameters Correlated**

| | | | | | |
|---|---|---|---|---|---|
| 1. $R_v$ | 5. $E_T^N$ | 9. $\beta$ | 13. $E_b$ | 17. $\chi_R$ | 21. SPP |
| 2. $Q_v$ | 6. $A_j$ | 10. AN | 14. $C_b$ | 18. $W$ | 22. SA |
| 3. $\beta_\mu^{1/2}$ | 7. $B_j$ | 11. DN | 15. $\pi^*$ | 19. $\mu$ | 23. SB |
| 4. $\delta_H$ | 8. $\alpha$ | 12. $A(^{14}N)$ | 16. $\pi_{azo}^*$ | 20. $\log K_{o/w}$ | |

For definitions and references, see Tables A.2a and b.

With this in view, statistical correlations have been undertaken by a number of authors, either pairwise (between two variables at a time) or between one, believed to be composite, and several others in multiple linear correlation, that is to say, seeking relationships of the form of Equation 4.1.

The 23 parameters listed in Table 4.1 (a selection of those listed in Tables A.2a and b) were subjected to linear correlation, two at a time. The results are displayed in Table 4.2: the binary correlation coefficients $r_{i,j}$ along with the numbers $n_{i,j}$ of solvents for which values of both parameters were available. Values of $|r_{i,j}|$ below 0.2 suggest that the parameters labeled $i$ and $j$ were essentially independent; values close to unity ($|r_{i,j}| > 0.9$) suggest that the two measure essentially the same property. Intermediate values suggest mixing of two or more. For instance, $E_T^N$, $\beta_\mu^{1/2}$, AN, and $W$ all show considerable cross-correlation. They appear in large part to measure similar properties.

To disentangle these cross-relationships, the statistical method of *principal component analysis* (PCA) (Eliasson *et al.*, 1982; Johnson, 1998; Malinowski and Howery, 1980; Thielemans and Massart, 1985) may be used. The purpose of PCA is to express the various properties in terms of a set of new variables, chosen to be independent of each other, that is to say, they are orthogonal. Any of the original variables can be written as a linear combination of the orthogonal set, just as a vector in three-dimensional space can be written in terms of three base vectors, mutually at right angles. Here, then, we write for each property $P_i$:

$$P_i = \sum a_{i,j} Z_j \qquad (4.2)$$

The number of the new variables, $Z_j$, is equal to the number of properties that were in the original set. It is a feature of PCA, however, that the principal components, $Z_j$, are oriented and ordered so that the first, $Z_1$ covers the largest part of the variance of the original data, $Z_2$ the next largest, and so on. In favorable cases, the number of new variables required to reproduce all the original data within experimental uncertainty, or at least to sufficient precision for their purpose, may be small. The sum in Equation 4.2, then, may contain only three or four significant terms, instead of the original number, here 23.

Parameters for which values were known for very few solvents, and solvents for which very few parameters were known, were omitted from consideration, as too many gaps vitiate the analysis. The data used, nevertheless, still contain many gaps.

**TABLE 4.2 Pairwise Correlation Coefficients Between Selected Solvent Parameters**

| No. | $R_v$ | $Q_v$ | $\beta_\mu^{1/2}$ | $\delta_H$ | $E_T^N$ | $A_j$ | $B_j$ | $\alpha$ | $\beta$ | AN | DN |
|---|---|---|---|---|---|---|---|---|---|---|---|
| | **1** | **2** | **3** | **4** | **5** | **6** | **7** | **8** | **9** | **10** | **11** |
| **1** | 1 | 28 | 28 | 22 | 28 | 27 | 27 | 27 | 27 | 22 | 17 |
| **2** | -0.491 | 1 | 28 | 22 | 28 | 27 | 27 | 27 | 27 | 22 | 17 |
| **3** | -0.348 | 0.896 | 1 | 22 | 28 | 27 | 27 | 27 | 27 | 22 | 17 |
| **4** | 0.063 | 0.734 | 0.807 | 1 | 22 | 21 | 21 | 21 | 21 | 17 | 16 |
| **5** | -0.480 | 0.848 | 0.687 | 0.876 | 1 | 22 | 27 | 27 | 27 | 22 | 17 |
| **6** | -0.463 | 0.705 | 0.482 | 0.739 | 0.950 | 1 | 27 | 26 | 26 | 21 | 16 |
| **7** | 0.061 | 0.660 | 0.793 | 0.805 | 0.493 | 0.328 | 1 | 26 | 27 | 21 | 16 |
| **8** | -0.492 | 0.585 | 0.311 | 0.445 | 0.889 | 0.937 | 0.085 | 1 | 26 | 21 | 16 |
| **9** | -0.321 | 0.632 | 0.490 | 0.230 | 0.504 | 0.371 | 0.306 | 0.387 | 1 | 21 | 16 |
| **10** | -0.430 | 0.637 | 0.377 | 0.751 | 0.949 | 0.983 | 0.178 | 0.942 | 0.346 | 1 | 15 |
| **11** | 0.262 | 0.046 | 0.107 | -0.115 | -0.053 | -0.149 | 0.105 | -0.143 | 0.769 | 0.037 | 1 |
| **12** | -0.444 | 0.677 | 0.546 | 0.689 | 0.938 | 0.969 | 0.453 | 0.895 | 0.407 | 0.960 | 0.163 |
| **13** | -0.173 | 0.567 | 0.586 | 0.257 | 0.249 | -0.055 | 0.465 | -0.137 | 0.794 | -0.165 | 0.057 |
| **14** | 0.147 | -0.236 | -0.175 | -0.381 | -0.402 | -0.375 | -0.301 | -0.276 | 0.397 | -0.314 | 0.901 |
| **15** | 0.107 | 0.705 | 0.744 | 0.795 | 0.649 | 0.532 | 0.942 | 0.316 | 0.341 | 0.460 | 0.098 |
| **16** | -0.022 | 0.788 | 0.825 | 0.813 | 0.696 | 0.556 | 0.922 | 0.352 | 0.437 | 0.539 | 0.133 |
| **17** | -0.062 | -0.821 | -0.723 | -0.846 | -0.860 | -0.825 | -0.741 | -0.584 | -0.462 | -0.747 | -0.111 |
| **18** | -0.168 | 0.883 | 0.551 | 0.983 | 0.962 | 0.903 | 0.369 | 0.786 | 0.152 | 0.923 | -0.003 |
| **19** | 0.594 | -0.177 | 0.002 | 0.045 | -0.379 | -0.439 | 0.200 | -0.474 | 0.060 | -0.282 | 0.585 |
| **20** | 0.051 | -0.581 | -0.598 | -0.744 | -0.676 | -0.622 | -0.544 | -0.478 | -0.482 | -0.576 | -0.154 |
| **21** | -0.086 | 0.918 | 0.890 | 0.832 | 0.744 | 0.597 | 0.867 | 0.390 | 0.540 | 0.514 | 0.172 |
| **22** | -0.475 | 0.506 | 0.171 | 0.729 | 0.939 | 0.962 | -0.109 | 0.924 | 0.266 | 0.945 | -0.047 |
| **23** | -0.170 | 0.430 | 0.302 | 0.139 | 0.220 | 0.110 | 0.125 | 0.135 | 0.904 | 0.052 | 0.848 |

*(Continued)*

**TABLE 4.2** (cont'd)

| No. | A($^{14}$N) | $E_b$ | $c_b$ | $\pi^*$ | $\pi^*_{azo}$ | $\chi_R$ | W | $\mu$ | Log $K_{o/w}$ | SPP | SA | SB |
|---|---|---|---|---|---|---|---|---|---|---|---|---|
| | **12** | **13** | **14** | **15** | **16** | **17** | **18** | **19** | **20** | **21** | **22** | **23** |
| 1 | 23 | 14 | 14 | 28 | 28 | 24 | 13 | 19 | 22 | 26 | 17 | 23 |
| 2 | 23 | 14 | 14 | 28 | 28 | 24 | 13 | 19 | 22 | 26 | 17 | 26 |
| 3 | 23 | 14 | 14 | 28 | 28 | 24 | 13 | 19 | 22 | 26 | 17 | 26 |
| 4 | 18 | 13 | 13 | 22 | 22 | 19 | 9 | 14 | 17 | 20 | 11 | 20 |
| 5 | 23 | 14 | 14 | 28 | 28 | 24 | 13 | 19 | 22 | 26 | 17 | 26 |
| 6 | 22 | 13 | 13 | 27 | 27 | 23 | 12 | 19 | 22 | 25 | 16 | 25 |
| 7 | 22 | 13 | 13 | 27 | 27 | 23 | 12 | 19 | 22 | 25 | 16 | 25 |
| 8 | 22 | 14 | 14 | 27 | 27 | 24 | 13 | 18 | 21 | 26 | 17 | 26 |
| 9 | 22 | 14 | 14 | 27 | 27 | 24 | 13 | 18 | 21 | 26 | 17 | 26 |
| 10 | 19 | 12 | 12 | 22 | 22 | 19 | 13 | 16 | 17 | 21 | 16 | 21 |
| 11 | 13 | 12 | 12 | 17 | 17 | 15 | 10 | 12 | 13 | 16 | 10 | 16 |
| 12 | 1 | 11 | 11 | 23 | 23 | 20 | 12 | 17 | 18 | 21 | 15 | 21 |
| 13 | 0.148 | 1 | 14 | 14 | 14 | 13 | 9 | 10 | 11 | 14 | 9 | 14 |
| 14 | −0.233 | 0.139 | 1 | 14 | 14 | 13 | 9 | 10 | 11 | 14 | 9 | 14 |
| 15 | 0.630 | 0.373 | −0.349 | 1 | 28 | 24 | 13 | 19 | 22 | 26 | 17 | 26 |
| 16 | 0.660 | 0.424 | −0.296 | 0.961 | 1 | 24 | 13 | 19 | 22 | 26 | 17 | 26 |
| 17 | −0.850 | −0.298 | 0.307 | −0.892 | −0.892 | 1 | 12 | 16 | 21 | 24 | 16 | 24 |
| 18 | 0.891 | −0.318 | −0.251 | 0.680 | 0.754 | −0.883 | 1 | 10 | 9 | 13 | 12 | 13 |
| 19 | −0.378 | −0.095 | 0.666 | 0.074 | 0.150 | 0.072 | −0.182 | 1 | 15 | 18 | 13 | 18 |
| 20 | −0.636 | 0.018 | 0.075 | −0.591 | −0.624 | 0.638 | −0.816 | 0.008 | 1 | 21 | 13 | 21 |
| 21 | 0.608 | 0.603 | −0.235 | 0.915 | 0.932 | −0.899 | 0.684 | 0.027 | −0.652 | 1 | 17 | 21 |
| 22 | 0.965 | −0.458 | −0.366 | 0.358 | 0.356 | −0.618 | 0.882 | −0.431 | −0.482 | 0.298 | 1 | 17 |
| 23 | 0.051 | 0.739 | 0.635 | 0.137 | 0.178 | −0.351 | −0.425 | 0.056 | −0.368 | 0.370 | −0.147 | 1 |

Entries above the diagonal represent the number of solvents $n_{i,j}$, for which both parameters were known. Entries on and below the diagonal are the correlation coefficients, defined by:

$$r_{i,j} = \frac{(n_{i,j}\Sigma x_i x_j - \Sigma x_i \Sigma x_j)}{[(n_{i,j}\Sigma x_i^2 - (\Sigma x_i)^2)(n_{i,j}\Sigma x_j^2 - (\Sigma x_j)^2)]^{1/2}}$$

Some of these exist merely because the necessary measurements have not yet been made, but in many cases the measurements are difficult or impossible, usually because of problems with solubility or chemical incompatibility. A body of data containing no gaps, when analyzed by PCA, should yield a set of positive numbers, representing the decreasing importance of successive components. These are the eigenvalues of the matrix of all the correlation coefficients, that is, the symmetric matrix obtained from Table 4.2 by replacing all the entries $n_{i,j}$ above the diagonal with $r_{i,j} = r_{j,i}$. A plot of these eigenvalues against the component number has a fancied resemblance to the pile of debris below a cliff, so is termed a scree plot. Such a plot for these data is shown in Figure 4.1. Eigenvalues less than unity are commonly considered to be negligible. Here apparently the first four PCs are significant. This is perhaps unfortunate, as it is difficult to represent more than two dimensions on the page. We shall contrive to represent the first three. The scree plot also shows the consequence of the gaps in the data in that the eigenvalues are not all positive. The last four have small negative values. These we shall ignore, though they contribute to the uncertainty.

## 4.3   REPRESENTATION OF CORRELATION DATA
## ON THE HEMISPHERE

The results of PCA of these correlations were applied to generate a graphical representation of the similarities and differences among the parameters correlated, proceeding as follows. The first three principal components, which are orthogonal by construction, are used to set up a coordinate system. Each of the $n$ original parameters $P_i$ ($i = 1, 2, \ldots, n$) is correlated with each of the three PCs $Z_r$ ($r = 1, 2, 3$). Then each of the original parameters is considered to be a vector in the space spanned by these coordinates, and its direction defined approximately by the angles it makes with the three axes. The cosine of each of these angles may be shown to be equal to the corresponding correlation coefficient. The uncertainty in direction arises from the part of the variance of the parameter associated with the fourth, fifth, and so on, principal components, from experimental error, and from gaps in the data. The magnitude of the vector does not affect the correlation coefficients, nor the angles (in fact, the magnitudes are lost in the correlation process), so the vectors can be treated as all terminating on the surface of a sphere of unit radius. In order to avoid having to show the whole surface of the sphere, it is convenient to plot as points on a hemisphere both those vectors that naturally appear there and the "antipodes" of those that would appear on the hidden side. (If two parameters have a large correlation coefficient, they measure the same property, even though the sign may be negative.)

In Figure 4.2, the 23 selected parameters are plotted in projection on a hemisphere. The first PC is labeled **N**, at the North Pole, and the second and third **G** (for Greenwich) and **E**,[2] on the Equator at 0 and 90° longitude, respectively. It

---

[2]Milne (1926, p. 133). "There's a South Pole," said Christopher Robin, "and I expect there's an East Pole and a West Pole, though people don't like talking about them."

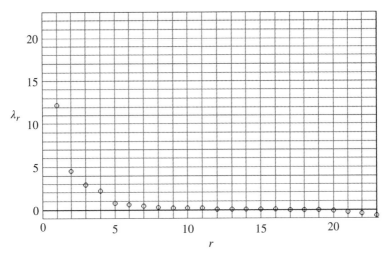

**FIGURE 4.1** Scree plot: eigenvalues $\lambda_r$ of the matrix of correlation coefficients of 23 parameters for 28 solvents, in descending order. Four eigenvalues are greater than unity, with a distinct break before the fifth, suggesting that four independent properties of the solvents are significant.

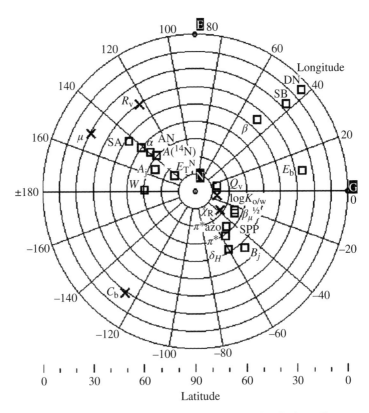

**FIGURE 4.2** Representation of the 23 parameters on the hemisphere. Squares represent positive vector directions, crosses negative, that is, antipodes. The Cartesian axes corresponding to the three principal components are labeled N(North pole), **G** (Greenwich meridian on the Equator), and **E** (90°E on the Equator).

may be shown that if the error is zero, the sum of the squares of the correlation coefficients that is, of the three direction cosines, is unity (Coxeter, 1961). Parameters that measure similar properties should appear close together. Measures that are essentially independent, that is to say, orthogonal, are nearly 90° apart, as the term implies.

Because the first PC (**N**) by construction contains the greatest part (53%) of the variance of all the data, many of the points are loosely scattered about the **N** axis. Some of the points are grouped as might be expected, but some unexpected similarities and differences appear. Notable groups are as follows:

1. The "acidic" group, clustered around latitude 50°, longitude 140°: SA, $\alpha$, AN, $A(^{14}N)$, and $A_j$.

2. The group spread from latitude 78°, longitude 13° to latitude 50°, longitude −60°, that measures various blends of (di)polarity and polarizability. The subgroup $Q_v$, $-\log K_{o/w}$, $-\chi_R$, $\beta_\mu^{1/2}$, and SPP appear to measure chiefly polarity, while $\pi^*$, $\pi_{azo}^*$, and $\delta_H$ contain increasing amounts of polarizability.[3] $R_v$, plotted as its antipodes at latitude 30°, longitude 123°, if plotted directly would appear just off the diagram at longitude −57°, extending the group to pure polarizability.

3. The "basic" group at latitudes 5–35°, longitude about 45°: DN, SB, and $\beta$, to which might be added $C_b$, if it is replotted directly, off the diagram at longitude 55°, not far from DN. $E_b$ is not unexpectedly off to one side, at latitude 26°, longitude 10. The "basity," $B_j$, lies at latitude 47°, longitude −48°, far from the other basic parameters. It appears to represent quite another property.

4. Two parameters related to polarizability, $R_v$ (latitude 31°, longitude 123°) and Marcus's softness parameter $\mu$ (latitude 20°, longitude 152°) appear fairly close together, plotted as antipodes.

5. $E_T^N$ lies between the acidic and polar groups. In an earlier version of this exercise (Buncel and Stairs, 2003, p. 72), Brownstein's $S$ and Kosower's $Z$ were almost coincident with $E_T^N$ and are omitted here. Winstein's log $k_1$, labeled $W$ (after Bartlett, 1972), is nearby.

Stairs and Buncel (2006) showed that one can use the three PCs to place a point representing the solvent effects on a reaction, expressed as the logarithm of either a rate constant or an equilibrium constant, on the hemisphere diagram constructed using solvent properties, and thereby show the relative importance of contributions by polarity/polarizability, acidity, and basicity of the solvent to the overall solvent effect. They showed, for instance, that in the *gauche–cis* equilibrium of chloroacetaldhyde (Karabatsos and Fenolio, 1969) while the more polar *cis* rotamer is favored

---

[3]In the previous edition (Buncel *et al.* 2003) we commented that the position of $\delta_H$ among the polarity/polarizability parameters might be due to the inclusion of the alcohols, with high but uncertain values. They were not included here.

by solvent polarity as expected, it is unexpectedly also favored by solvent basicity. Apparently chloroacetaldehyde is more acidic in the *cis* than in the *gauche* form. Similar treatment of kinetic data showed that the rate of epoxidation of cyclohexene by peroxybenzoic acid is considerably accelerated by both acidity and polarity of the solvent, but it is retarded by solvent basicity, owing to interaction of the solvent with the peroxybenzoic acid.

Stairs and Buncel (2006) were discussing use of the method to survey a number of reactions in one study, and cast a wide net (with some holes in it). In applying the method to analysis of a single reaction or a group of related reactions, it would be preferable to carry out the measurements of rate or equilibrium in a set of solvents covering a range of each of the properties acidity, basicity, polarity, and polarizability, however expressed, setting up the frame by performing PCA on these properties, and then placing the reaction data in the frame. That way the statistical problems arising from gaps in the data would be avoided, and the results would be easier to interpret. An example of this approach is based on the work of Romeo *et al.* (1980) on three reactions of *cis*–[Pt(PEt$_3$)($m$–C$_6$H$_4$CH$_3$)Cl], an uncharged complex, which exhibits square-planar coordination about Pt. The reactions were as follows:

1. Uncatalyzed (spontaneous) isomerization *cis* → *trans*; rate constant $k_i$.
2. Displacement of Cl$^-$ by a solvent molecule; rate constant $k_1$.
3. Displacement of Cl$^-$ by thiourea; rate constant $k_2$.

Table 4.3 lists the observed rate constants (as common logarithms) for the three reactions in the eight solvents, together with the values of two acid parameters ($\alpha$ and SA), two basic parameters ($\beta$ and SB), and three considered to measure polarity ($E_T^N$, $\pi^*$, and SPP). The last two rows list the coordinates of the 10 points on the hemisphere. They are shown in Figure 4.3. It is apparent that two of the reactions are accelerated by solvent acidity: the isomerization ($k_i$) strongly and the replacement of Cl$^-$ by solvent ($k_1$) weakly; the two acid parameters and $E_T^N$ (which has a large acid component) and these two rate constants lie in a close cluster. (They are all shown as "antipodes.") They are essentially unaffected by solvent polarity ($\pi^*$ and SPP) and by solvent basicity ($\beta$ and SB) all of which lie about 90° away (orthogonal). The displacement of Cl$^-$ by thiourea, on the other hand, is independent of solvent acidity, weakly accelerated by solvent basicity, and if anything retarded by polarity ($\pi^*$ and SPP, nearby, are plotted directly, while $k_i$ is antipodal).

It is possible to construct vectors in the space spanned by the three PCs that are parallel to the vectors representing the reactions in this space. Figure 4.3 show plots of the logarithms of each reaction constant versus the corresponding elements of the appropriate parallel vector. The slopes of these plots show the sensitivity of the reactions to solvent change. It is apparent that the isomerization reaction is nearly three times as sensitive to solvent change as are the two replacement reactions (Fig. 4.4).

**TABLE 4.3  Rate Constants of Three Reactions of A Square-planar Pt(II) Complex in Eight Solvents, with Values of Seven Parameters for Each Solvent[a]**

| Solvent | $\log k_i$ | $\log k_1$ | $\log k_2$ | $E_T^N$ | $\alpha$ | $\beta$ | $\pi^*$ | SA | SB | SPP |
|---|---|---|---|---|---|---|---|---|---|---|
| Methanol | -1.73 | 0.47 | 4.12 | 0.762 | 0.98 | 0.66 | 0.60 | 0.605 | 0.545 | 0.857 |
| Ethanol | -2.99 | 0.29 | 4.28 | 0.654 | 0.86 | 0.75 | 0.54 | 0.400 | 0.658 | 0.853 |
| 1-Propanol | -3.44 | 0.19 | 4.50 | 0.617 | 0.84 | 0.90 | 0.52 | 0.367 | 0.727 | 0.847 |
| 1-Butanol | -3.79 | 0.20 | 4.67 | 0.586 | 0.84 | 0.84 | 0.47 | 0.341 | 0.809 | 0.837 |
| 2-Methoxyethanol | -3.92 | -0.34 | 4.01 | 0.657 | (0.76) | (0.66) | (0.54) | 0.355 | 0.560 | 0.822 |
| 2-Propanol | -4.04 | -0.18 | 4.65 | 0.546 | 0.76 | 0.84 | 0.48 | 0.283 | 0.762 | 0.848 |
| t-Butanol | -5.69 | -0.94 | 4.83 | 0.386 | 0.42 | 0.93 | 0.41 | 0.145 | 0.928 | 0.829 |
| Acetonitrile | -5.38 | -0.69 | 3.66 | 0.460 | 0.19 | 0.40 | 0.66 | 0.044 | 0.286 | 0.895 |
| *Latitude* | *13.9* | *18.2* | *60.7* | *8.8* | *38.6* | *70.1* | *60.2* | *25.8* | *67.1* | *60.7* |
| *Longitude* | *-168.2* | *-155.9* | *-38.3* | *176.4* | *-171.0* | *-45.2* | *4.5* | *-173.9* | *-24.2* | *83.9* |
| Sensitivity | 3.33 | 1.28 | 1.04 | | | | | | | |

[a] Data from Romeo *et al.* (1980).

The last three rows are the coordinates obtained by PCA of the seven parameters, as represented in hemispheric projection, the coordinates of the reaction rate constants and in the same frame, and the slope of the plot of log *k* versus the "parallel" vector elements (see text).

Values of parameters in parentheses for 2-methoxyethanol are estimates.

Coordinates in italics are for the antipodes of points that would lie on the hidden hemisphere.

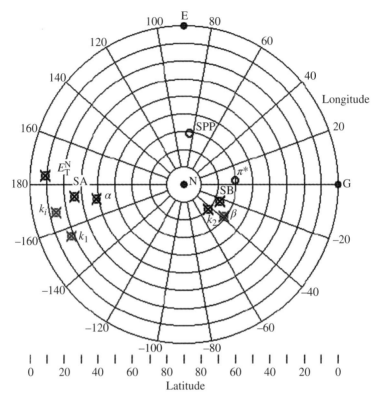

**FIGURE 4.3** Hemisphere projection: three rate constants and seven solvent parameters. Circles represent points in their natural positions, crossed circles antipodes of points that fall in the hidden hemisphere, that is, the negatives of the indicated variables. (*See insert for color representation of the figure.*)

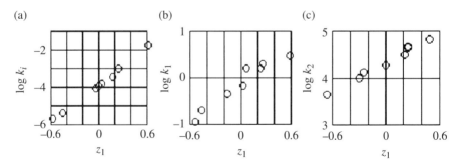

**FIGURE 4.4** Rate constants (logarithms) for reactions, (a–c) in the seven solvents versus the corresponding elements of the parallel vectors $z_i$.

## 4.4   SOME PARTICULAR CASES

Beginning in 1972, Kamlet, Taft, and coworkers carried out a long series of studies[4] of solvent effects, beginning with the effects on spectra (frequency shifts and intensities). They show that it is necessary to consider not only the bulk polarity of the solvent, but also the ability of the solvent to act as donor or acceptor in hydrogen bonding. They use the parameter $\pi^*$ to represent the bulk polarity, choosing this symbol because it is based on frequency shifts of $\pi \to \pi^*$ and $n \to \pi^*$ electronic transitions of selected dyes. The hydrogen-bonding ability is represented by two parameters: $\alpha$ for the solvent's tendency to act as a proton donor (acidic character) and $\beta$ for its proton acceptor tendency (basic character). They then write Equation 4.3 (*cf* Eq. 4.1):

$$XYZ = XYZ_0 + s\pi^* + a\alpha + b\beta \tag{4.3}$$

where the symbol $XYZ$ represents the property being studied (expressed in such a way that it is proportional to an energy, as, for example, a spectral frequency or the logarithm of a rate or equilibrium constant), the subscript 0 indicates its value in the reference solvent; $\pi^*$, $\alpha$, and $\beta$ are the solvent parameters defined earlier; and $s$, $a$, and $b$ are the sensitivity coefficients for the property $XYZ$ to the corresponding solvent property. Values of $\pi^*$, $\alpha$, and $\beta$ are listed for some solvents in Table A.3.

For an example of a pairwise correlation, consider the following two quantities that purport to measure the polarity of the solvent. Kosower (1958, 1968) proposed a parameter, $Z$, based on the change in frequency of a charge-transfer transition of 1-ethyl-4-methoxycarbonyl-pyridinium iodide (**1**), which shows a pronounced increase of frequency when the polarity of the solvent is increased, because the ionic ground state is stabilized by a polar medium relative to the nonionic excited state. Dimroth and Reichardt (1969; Reichardt and Harbusch-Görnert, 1983) proposed another, $E_T(30)$, or in "normalized" form, $E_T^N$, based on a visible transition of a pyridinium-*N*-phenoxide betaine dye (**8**, p, 40). (The label (30) is because the dye was the 30th of a number tried.) Figure 4.5 shows the correlation of these two parameters for 40 solvents.

The correlation does not appear to be very good, but in fact, by using a select set of solvents believed to be free from interferences such as specific interactions or purity problems, it has been used to obtain estimates of $E_T(30)$ for solvents in which the dye is insoluble or reacts. With these additions $E_T(30)$ became available for 271 solvents (Reichardt, 1994).

Another example with a different result was the attempted correlation of the theoretically justified quantity $\beta_\mu$, defined by Dutkiewicz (1990) as $(2N_A^2/3\varepsilon_0)(\mu^2/V)$ (proportional to the square of the dipole moment of the solvent molecule divided by its molar volume[5]), with $E_T^N$. She found that, instead of lying close to a single line, the points fell into four fairly distinct classes (with a few outliers). See Figure 4.6.

---

[4]For example, Abboud and Taft (1979), Kamlet and Taft (1979), Kamlet *et al.* (1979), Kamlet *et al.* (1981), Taft *et al.* (1981).
[5]The symbol $\beta_\mu$, is used here to distinguish it from the Taft–Kamlet basic parameter $\beta$.

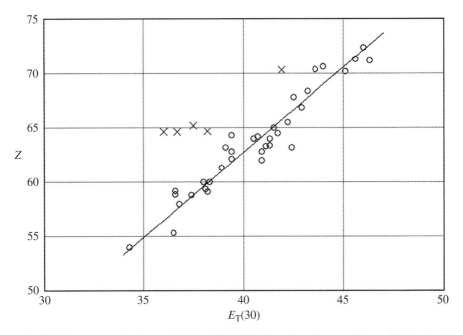

**FIGURE 4.5** Kosower's $Z$ versus Reichardt's $E_T(30)$. Data from the compilation of Abboud and Notario (1999). Points marked with crosses are omitted from the statistics. They are for compounds flagged by the last authors as uncertain and a group of esters and ethers that seem to have values of $Z$ clustered about 64–65 kcal mol⁻¹, lying above the trend. No strong HBD solvents are included.

The main classes were as follows:

1. Weakly dipolar and nonpolar, non-HBD (hydrogen-bond donor) molecules: hydrocarbons and halogenated derivatives, ethers, esters, tertiary amines.
2. Strongly dipolar, non-HBD: ketones, $N,N$-dialkyamides, nitro-compounds, sulfoxides, sulfones, pyridine.
3. HBD molecules: water, primary alcohols, glycols, carboxylic acids, with nonprimary alcohols, aniline, and perhaps chloroform (weak HBDs) forming subclass 3a, and phenols (strong HBDs), 3b.
4. $N$-monoalkylamides and formamide (strongly dipolar, weak HBDs).

The distinction among these classes seems to rest mainly upon their ability to function as donors in the formation of hydrogen bonds (HBD). To test this, we tried a bilinear correlation of $E_T^N$ with $\beta_\mu$ and $\alpha$, derived by Kamlet and Taft (discussed earlier), which is taken as a measure of HBD strength. The results may be expressed by Equation 4.4, with the constants $C_0 = 0.1501$, $C_1 = 0.003764$, and $C_2 = 0.4974$. The standard error of estimate was 0.05.

$$E_T^N = C_0 + C_1\beta_\mu + C_2\alpha \tag{4.4}$$

Figure 4.7 shows a plot of $E_T^N$ versus the composite variable $(C_1\beta_\mu + C_2\alpha)$. A plot of $(E_T^N - C_2\alpha)$ versus $\beta_\mu$, expected to be linear, showed some curvature; we therefore

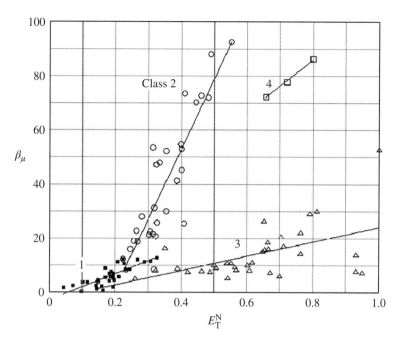

**FIGURE 4.6** The dielectric parameter $\beta_\mu$ plotted against $E_T^N$ for 101 solvents. Subclasses 3a, 3b, and chloroform are combined as class 3. Source: Dutkiewicz (1990) with permission of the Chemical Society.

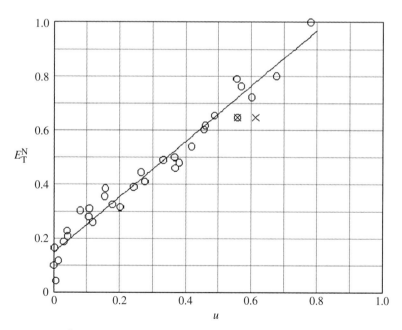

**FIGURE 4.7** $E_T^N$ versus the composite variable: $u = C_1\beta_\mu + C_2\alpha$; $C_1 = 0.003764$, $C_2 = 0.4974$. In this and the following figure, the points marked by crosses are for acetic acid. The circled cross is for the dimer; the plain cross is for the monomer. Neither is included in the statistics. The standard error of estimate $s = 0.046$.

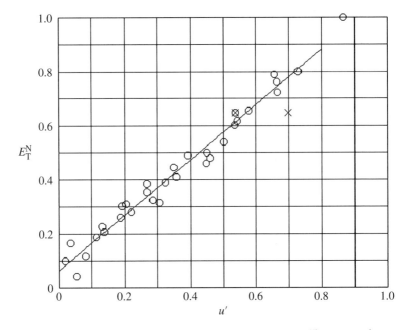

**FIGURE 4.8** $E_T^N$ versus the new composite variable: $u' = C_1'\beta_\mu^{1/2} + C_2'\alpha$; $C_1' = 0.04174$, $C_2' = 0.4792$, $s' = 0.037$. See caption for Figure 4.5.

tried $\sqrt{\beta_\mu}$. Figure 4.8 shows the result of plotting $E_T^N$ versus the new composite variable $(C_1'\beta_\mu^{1/2} + C_2'\alpha)$. The fit is not notably better, judged by the standard error, but the linearity is improved, especially near the origin. Since $\alpha$-values are not available for all the solvents in Dutkiewicz's classification there are fewer points plotted, but we suggest (without any theoretical justification) that the square root of $\beta_\mu$ is a better measure of the contribution of the dipole moment to the bulk polarity than $\beta_\mu$ itself.

Figure 4.9 is similar to Figure 4.6, except that the square root of $\beta_\mu$ is now the ordinate. With this change, it appears that classes 1 and 2 are no longer clearly differentiated, but are scattered about one line. Since they are similar in lacking HBD ability, and differ chiefly in dipolarity, this is not surprising.

Carboxylic acids present a problem in assigning $\beta_\mu$. Acetic acid, for instance, associates strongly through hydrogen bonding to form the symmetrical dimer (2) in the pure liquid or in solution in nonhydroxylic solvents. Should the value of the dipole moment, $\mu$, be that of the monomer (1.74 Debye or $5.6 \times 10^{-30}$ cm, in the gas phase), or that of the dimer, which is zero? Perhaps it should be assigned some intermediate, probably variable, value representing the effect of the few free monomers or the tendency of the dimer to dissociate in the presence of a polar solute molecule with which it can interact. In each of Figures 4.5 and 4.6 acetic acid appears twice, marked by crosses: once with $\beta_\mu = 14.97 \, \text{kJ mol}^{-1}$ (plain cross) and once with zero (circled cross). In both figures it is the point for the dimer that seems closest to the trend of the other points.

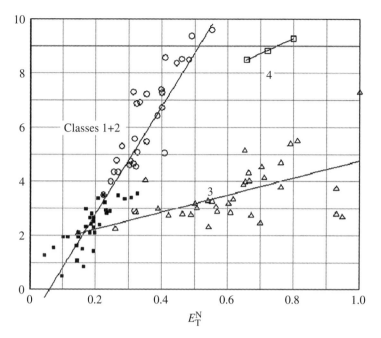

**FIGURE 4.9**    Similar to Figure 4.6, but plotting the square root of $\beta_\mu$. Classes 1 and 2 are now combined.

Swain *et al.* (1983) have proposed a pair of parameters, $A_j$ and $B_j$, called *acity* and *basity*, coinages to imply similarity to acidity and basicity, but properties of the bulk solvents rather than the molecules. They fitted data for a great many phenomena (reaction rates, equilibria, spectral shifts) in a variety of solvents to equations of the form of Equation 4.5:

$$p_{i,j} = c_i + a_i A_j + b_i B_j \qquad (4.5)$$

where $p_{i,j}$ and $c_i$ have the dimensions of energy. Four fixed points served to anchor the scales. These were $A_j = B_j = 1$ for water and $A_j = B_j = 0$ for *n*-heptane and two additional conditions of $A_j = 0$ for hexamethylphosphoric triamide and $B_j = 0$ for trifluoroacetic acid.

Acity represents some composite of all the properties of a substance that are associated with positive charge: HBD activity (tendency to donate a shared proton in a hydrogen bond), Brønsted acidity (tendency to donate a proton outright), Lewis acidity (tendency to receive a shared electron pair), electron affinity (tendency to receive an electron outright), oxidizing strength, electrophilicity; in a phrase, Usanovich acidity. Basity similarly represents a composite of all the properties that are associated with negative charge: Brønsted or Lewis basicity (tendency to donate a shared pair to a proton or to a vacant orbital, such as on a metal ion), the negative

TABLE 4.4    Coefficients of Correlation Between $E_T^N$
(1) and $A_j$ (3) and $B_j$ (4) and with Either $B_\mu$ or $\beta_\mu^{1/2}$ (2)

| Dependent variables | $r_{1,2}^2$ | $r_{1,3}^2$ | $r_{1,4}^2$ |
|---|---|---|---|
| $\beta_\mu, A_j, B_j$ | 0.544 | 0.940 | 0.178 |
| $\beta_\mu^{1/2}, A_j, B_j$ | 0.731 | 0.947 | 0.039 |

of ionization potential (tendency to lose an electron outright), reducing strength, and nucleophilicity; in sum, Usanovich basicity.

We have correlated $E_T^N$ with $A_j$ and $B_j$ and with $\beta_\mu$ both directly and as its square root. Table 4.4 shows the coefficients of correlation for these data. Again, the square root of $\beta_\mu$ gives an improved correlation, and acetic acid fits the trend better if assigned $\beta_\mu =$ zero. It appears that in bulk acetic acid the dimer is stable enough to cause the liquid to act as a nonpolar medium, but labile enough to allow interaction with hydrogen-bond acceptor (HBA) molecules. It may be noted that when the square root of $\beta_\mu$ is used, the correlation coefficient with $B_j$ becomes 0.039, which is effectively zero. In all the foregoing correlations, $E_T^N$ turns out to be as closely related to the acidity (in the guise of HBD strength or acity) as to the (di)polarity. It is essentially uncorrelated with polarizability, as measured by the polarization function: $R_v = (n^2 - 1)/(n^2 + 2)$.

## 4.5   ACIDITY AND BASICITY PARAMETERS

The two ac(id)ity parameters $\alpha$ and $A_j$ are strongly correlated ($r_{6,8} = 0.937$ using a set of 26 solvents for which both quantities are known). Within the set, though there seems to be a difference in behavior between C-acids (chlorohydrocarbons, ketones, nitromethane and acetonitrile) on the one hand and O-acids (alcohols, water, acetic acid) on the other. The slopes obtained when the values of $\alpha$ are plotted versus $A_j$ for the O-acids and for the C-acids (Fig. 4.10) are probably the same within experimental error ($1.0 \pm 0.2$), but the points for the latter are displaced about 0.2 units of $\alpha$ downward. With caution, the correlations may allow estimation of $\alpha$ where values are not known.

Catalán (2001) records values of three new parameters for 190 solvents. They are SPP, believed to be a measure of pure polarity and polarizability uncontaminated by acidity, SB, a measure of basicity, and SA, a measure of acidity. SPP is based on the difference $\Delta\nu$ between the frequencies of the first absorption maxima of 2-($N,N$-dimethylamino)-7-nitrofluorene (DMNAF, 3) and 2-fluoro-7-nitrofluorene (FNF, 4), according to Equation 4.4:

$$SPP = \frac{\Delta\nu(\text{solvent}) - \Delta\nu(\text{gas})}{\Delta\nu(\text{DMSO}) - \Delta\nu(\text{gas})} \tag{4.6}$$

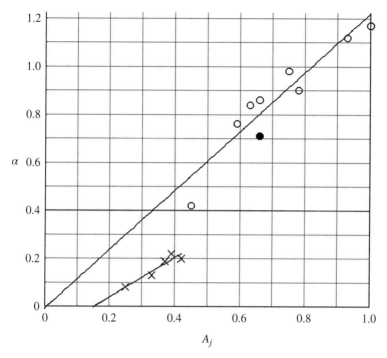

**FIGURE 4.10**    Acidity parameters: Taft–Kamlet $\alpha$ versus Swain's $A_j$. Circles are for O-acids and (filled) formamide; crosses are for C-acids. Separate least-squares lines are fitted for O-acids and for C-acids. Data from Kamlet *et al.* (1983) and Swain *et al.* (1983), with corrections Reichardt and Welton (2011).

SB is similarly based on the difference in frequency between 5-nitroindoline (NI, **5**) and 1-methyl-5-indoline (MNI, **6**), according to Equation 4.7. (TMG is tetramethylguanidine.)

$$SB = \frac{\Delta v(\text{solvent}) - \Delta v(\text{gas})}{\Delta v(\text{TMG}) - \Delta v(\text{gas})} \tag{4.7}$$

The acid parameter, SA, is based first on the relation between the observed frequencies, measured in nonacidic solvents, of the stilbazolium betaine dye **7** (TBSB: R=H) and the $O,O'$-di-$t$-butyl derivative (DTBSB: R = $t$-butyl) represented by Equation 4.8.

$$v_{\text{TBSB}} = 1.409 v_{\text{DTBSB}} - 6288.7 \tag{4.8}$$

Then, for acid solvents,

$$SA = 0.400 \frac{\left(v_{\text{TBSB}} - (1.409 v_{\text{DTBSB}} - 6288.7)\right)}{1299.8} \tag{4.9}$$

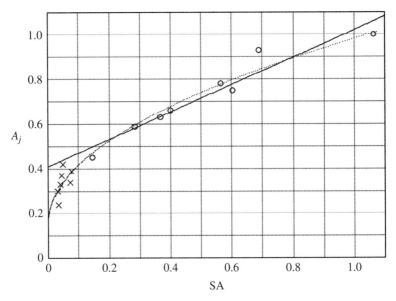

**FIGURE 4.11** Swain's acity parameter $A_j$ versus Catalán's (2001) SA. Circles are for O-acids (alcohols and acetic acid); crosses are for assorted C-acids. The straight line, fitted to the O-acids only, is represented by $A_j = 0.41 + 0.61$ SA; the curve, fitted to all, is represented by $A_j = 0.16 + 0.83\sqrt{SA}$.

which fixes the value for ethanol at 0.400. SA shares with $\alpha$ the assignment of zero to a number of solvents that have nonzero values of $A_j$. The two acidity parameters $A_j$ and SA are compared in Figure 4.11, plotting only those points for which both measures are nonzero. There is a reasonable correlation between them for the O-acids, but the remainder of the points cluster about a line with nearly infinite slope. A similar plot of $\alpha$ against SA gave similar results, the alcohols following a linear trend with slope 0.55 (with the exception of 2,2,2-trifluoroethanol) and the C-acids in a loose cluster.

As noted earlier, there are also some differences among the basic parameters. We considered nine, varying in their emphasis of HBA (hard) or Lewis (hard or soft) basicity. Those that would be expected to be in the hard range were $\beta$ (Kamlet and Taft, 1976) and Catalan's (2001) SB, and Hahn *et al.* (1985) and $\Delta\delta$ ($CHCl_3$), which all involve hydrogen bonding, and Maria *et al.*'s (1987) $-\Delta H(BF_3)$; $BF_3$ is a hard acid. Gutmann's (1967) donor number, DN, measures interaction with the borderline acid $SbCl_5$. $D_s$ (Persson *et al.* 1987) is designed to be soft, as it depends on interaction with the soft Lewis acid $HgBr_2$. The pair $E_b$ and $C_b$ was designed by Drago (1980) and coworkers to represent the electrostatic and covalent parts, respectively, of the enthalpy of reaction of the base with a Lewis acid. According to Klopman's (1968) analysis, hardness should be represented by a high $E_b$ relative to $C_b$ and softness by the reverse. The nature of Swain *et al.* (1983) $B_j$ is difficult to characterize, as it is based on correlation of a great many properties.

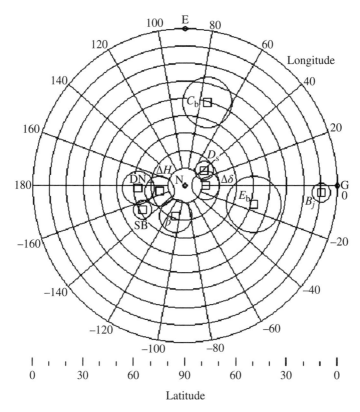

**FIGURE 4.12** Nine basic parameters for 13 non-HBD solvents. Parameters represented are (latitude, longitude in degrees): 1. $\beta$ (72, −107); 2. $B_j$ (9, −3); 3. SB (62, −151); 4. DN (63, −177); 5. $C_b$ (41, 75); 6. $E_b$ (48, −15); 7. $-\Delta H_{BF3}$ (75, −168); 8. $D_s$ (76, 36); 9. $\Delta\delta(CHCl_3)$ (78, 0) The error circles are somewhat arbitrary; they are drawn with radii proportional to the difference between the sum of squares of the three direction cosines and unity.

Figure 4.12 shows how the 9 parameters, for 13 non-HBD solvents, are disposed in the frame of the first three PCs. The three largest eigenvalues of the matrix of correlation coefficients were 6.53, 1.58, and 0.87, the rest much smaller. Three supposedly hard parameters $\beta$, SB, and $-\Delta H(BF_3)$ appear together, with $\Delta\delta(CHCl_3)$ and $E_b$ at a distance. All are far from $C_b$, as would be expected. $B_j$ again is off by itself (see Section 4.3). It is nearer $E_b$ than any other, suggesting that it measures mainly hard interactions, but not in the same manner as the other hard measures. This analysis reinforces the idea that "basicity" is not a simple property.

Laurence and coworkers (Laurence *et al.*, 2014) emphasize the distinction between basicity parameters of bulk solvents and those derived from measurements on substances in solution in inert solvents. In the same paper, they discuss some difficulties in the assignment of basicity parameters based on solvatochromism, such as the Taft–Kamlet $\beta$, to amphiprotic solvents, and propose another measure of H-bond basicity, $\beta_1$, based on the difference between the [19]F NMR chemical shifts of 4-fluorophenol and

4-fluoroanisole, which for non-HBD solvents is well correlated with $\beta$, but which should be free from complications due to HBA effects on the probe molecules used in the solvatochromic method. Values of $\beta_1$ for more than 70 solvents are tabulated, plus 40 "secondary" values, estimated by from correlations with selected solvatochromic shifts.

## 4.6    BASE SOFTNESS PARAMETERS

Chen *et al.* (2000) considered the relationships among three solvent softness scales relevant to bases, namely, Marcus's $\mu$ (discussed in Chapter 3), $D_s$ (discussed earlier) and a third measure, symbolized by $\Delta\Delta\nu(\text{I–C})$, which is the shift of the C–I stretch frequency if the molecule ICN on complexing with an EPD molecule in $CCl_4$ solution, corrected for a supposed hard component by subtraction of a small constant times the shift of the O–H stretch of phenol similarly treated. To these we added

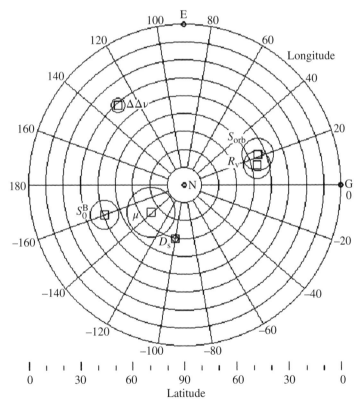

**FIGURE 4.13**    Six parameters for base softness. Three ($\mu$, $D_s$, and $\Delta\Delta\nu(\text{C–I})$) are discussed by Chen *et al.* (2000). $R_v$ is the volume polarization, $S_0^B$ is derived from the Drago parameters $C_B$ and $E_B$, and $S_{orb}$ is the reciprocal of the LUMO–HOMO energy difference. Values from Mu *et al.* (1998).

the volume polarization $R_v$, (for softness is conceptually related to polarizability), $S_0^B = 1.35 - E_B + 0.109C_B$ suggested by Figure 3.10b, and $S_{orb}$, the reciprocal of the LUMO–HOMO energy difference, suggested by Klopman (1968; Huheey *et al.*, 1993, p. 351) as further measures of softness.

Figure 4.13 shows the results of PCA on these 6, with a limited set of 15 non-HBD solvents for which most data were available for all 6, on a hemisphere projection using a new, appropriate set of coordinates. The first three eigenvalues corresponded to 54, 26, and 13% of the total variance of the data. The points fall into two clusters, with $S_{orb}$ and $R_v$ very close on one side, and $\mu$ and $D_s$ fairly close on the other. $\Delta\Delta\nu(I–C)$ lies at a distance from both clusters. Chen *et al.* (2000) proposed using correlations between $\mu$ and $D_s$ and between $\mu$ and $\Delta\Delta\nu(I–C)$ to estimate values of $\mu$ for solvents in which it could not be measured, for reasons including poor solubility. This picture suggests that $D_s$ is the better choice, though they were able to find values in approximate agreement using both methods, with a wider range of solvents.

The separation of the two clusters may be because $S_{orb}$ and $R_v$ are properties of the whole molecule, whereas $\mu$, $D_s$, and $S_0^B$ relate specifically to the coordination site. Why $\Delta\Delta\nu(I–C)$ lies so far from both clusters (nearly orthogonally) is not clear (but ICN is a rather strange sort of acid). Apparently base softness, like basicity itself, is not a simple property.

## 4.7  CONCLUSION

The statistical methods of chemometrics, particularly principal component analysis, have been helpful in sorting out confusing data in many aspects of chemistry, and linear free energy relationships in particular have been fruitful in providing insight into reaction mechanisms. The calculations discussed in this chapter only begin to show what could be done to aid understanding of some of the empirical parameters applied to the study of solvent effects. These techniques afford views of reactions that resemble impressionist paintings. For full understanding of the way in which reactions occur, we must wait until theories of elementary reactions (reaction dynamics) and of the structure and energetics of solvents and solvates have reached suitable levels. Progress in these matters is now rapid. In the next chapter we hope to show the direction in which the theory of solutions is moving, and to provide a brief introduction to the ways in which some theoretical techniques can be applied to problems in the field of reactions in solution.

# 5

# THEORIES OF SOLVENT EFFECTS

## 5.1 INTRODUCTION: MODELING

The purpose of computer-based modeling of chemical systems is to mimic them in properties of interest and to understand at a molecular or even an electronic level the physical origins of these properties. The ultimate goal in modeling a reaction in solution is to show, by modeling the interaction of dissolved reactants, activated complexes, and products with solvent molecules, how changing the solvent affects the rate or equilibrium of a reaction. In some cases, we may learn why a solvent may favor one of a set of possible mechanisms. To study the solvation of a single solute species, we may consider a two-stage model representing, first, an assembly of solvent molecules plus an isolated solute molecule or ion and, second, the same solute now surrounded by the solvent. When suitably translated into molar quantities, the changes in $U$, $H$, $S$, $G$, and $\gamma$ then yield estimates of the transfer energy, enthalpy, entropy, Gibbs energy, and activity coefficients of the solute from the ideal gas phase to the dilute solution. Though in practice this goal is not generally attainable, the calculations often provide valuable insight.

Of the available methods, *quantum mechanics* (QM) attacks the problem at its deepest level. Moore (1972), in one edition of his physical chemistry text, says that, in principle, all of chemistry could be calculated from the Schrödinger equation. Then in a footnote, he adds: " 'In principle' from the French, '*En principe, oui*', which means, '*Non*.' " Since that date, however, computers and programs have become more powerful, and much effort is being made to carry out quantum-mechanical calculations of the energetics of solvation of molecules and ions in various solvents. QM calculations are implemented in ab initio form at various levels of approximation, semiempirically also at various levels, and through density-functional theory.

*Solvent Effects in Chemistry*, Second Edition. Erwin Buncel and Robert A. Stairs.
© 2016 John Wiley & Sons, Inc. Published 2016 by John Wiley & Sons, Inc.

Other methods treat the molecules as classical objects, interacting through forces of various kinds and using the formalism of statistical thermodynamics to obtain the desired quantities as averages. The chief of these methods are *Monte Carlo* (MC), so called because it depends on (simulated) chance in the form of random numbers in evaluating average properties, and *molecular dynamics* (MD), which uses the laws of motion to explicitly represent the evolution in time of an assembly of molecules. The desired properties are obtained as time averages. There are also various hybrid methods, in which, for instance, the translation and rotation of molecules are assumed to behave classically but the internal vibrations are treated quantum mechanically or the solute is treated quantum mechanically but the solvent molecules are treated classically, and methods in which solute molecules are treated as neither in vacuo nor surrounded by other molecules, but as in a cavity in a continuous dielectric.

Finally, there is a group of methods labeled *integral equation theories*. These begin with the pairwise potential energy function, which may be derived from quantum calculation or may take one of several forms that are more or less realistic. The simplest is the "hard sphere" potential, which is zero for separations greater than an assumed collision distance and infinite within that distance. Other forms of potential are mentioned in Section 5.3 (see also Berry *et al.* (2000, p. 305)). The radial distribution function, $g(\mathbf{r})$, measures the departure from the mean of the number density of molecules around one molecule taken as fixed at the origin, as a function of the radius vector $\mathbf{r}$. The integral equation theories are so called because they give rise to equations in which $g(\mathbf{r})$ appears both inside and outside an integral over another radius vector, $\mathbf{r'}$.

We shall not attempt to describe any of these methods in detail, but offer outlines of them and take a look at some particular results.

## 5.2   QUANTUM-MECHANICAL METHODS

### 5.2.1   Ab Initio Methods

Quantum-mechanical calculations on single molecules can be done at different levels of theory. The most fundamental approach, the so-called "ab initio" ("from the beginning") method (Hehre *et al.*, 1986), begins with a guess at the structure of a molecule, guided by experience. The *Schrödinger equation* (SE) can be solved exactly for hydrogen-like atoms (a nucleus and one electron only). It can be solved nearly exactly for the hydrogen molecule ion $H_2^+$, subject only to the *Born–Oppenheimer approximation*: since nuclear motions are three orders of magnitude slower than electronic motions, the positions of the nuclei are assumed to be fixed, and the electrons treated as moving in the field created by the fixed nuclei. For other not-too-large atoms and for small molecules, the SE can be solved to arbitrary closeness of approximation, with due effort. For larger systems, the effort, that is to say the requirement of computer time and memory, grows rapidly with increasing numbers of electrons and nuclei, so lower levels of approximation become necessary. This is generally true whether the system consists of a single large molecule or more than a very few smaller molecules in interaction, though certain methods have been applied successfully to large systems.

The Born–Oppenheimer approximation is universally applied in calculations relevant to solvation. Electrons are accommodated in molecular orbitals, which are represented as *linear combinations of atomic orbitals* (LCAO). The atomic orbitals chosen form the *basis set*. One then calculates the energies of electrons in these molecular orbitals. The energy of the molecule is the sum of all the electron orbital energies, which are negative, plus the positive energies of mutual repulsion of all the nuclei. Repetition of the whole calculation with slightly shifted nuclei enables one to find the configuration of minimum energy and the shape of the hypersurface representing the energy as a function of nuclear coordinates in the vicinity of the minimum. As far as nuclear motions are concerned, this hypersurface represents what is effectively potential energy. From it, the vibrational frequencies and the *zero-point energies* associated with them can be obtained. The energy of the molecule at $0\,K$ is the energy minimum plus the zero-point vibrational energies.

In the simplest method, the *Hartree–Fock* (HF) or *self-consistent field* (SCF) method, each electron is treated as if it moved in an average electric field due to the nuclei and all the other electrons. The calculations may be done, in the simplest version, by taking as the basis set just enough atomic orbitals to represent the core and valence shells of each atom: for example, for the water molecule, the $1s$, $2s$, $2p_x$, $2p_y$, and $2p_z$ on oxygen and a $1s$ on each hydrogen, seven in all. These seven can be combined, observing the symmetry of the molecule, to form seven molecular orbitals. This is the *minimum basis set*. The orbitals themselves may be of the *Slater type*, which are simplified versions of hydrogen-like orbitals, with the oscillations in the radial part of the functions smoothed out. In current practice, they are almost always of the *Gaussian type*: approximation of Slater or hydrogen-like orbitals by linear combinations of several Gaussians (see Fig. 5.1). This greatly simplifies the calculations chiefly because the product of two Gaussians is another Gaussian (Fig. 5.2).

The light curves are the Gaussians, and the dotted curve is their weighted sum. Note that in each case the magnitude of the orbital is underestimated near the nucleus.

Gaussian approximations necessarily have two failings: they underestimate the magnitude of the wave function both at $r=0$ (underestimating both charge and spin density close to the nucleus) and at large $r$ (underestimating interatomic interactions through overlap). Fewer than three Gaussians per orbital make these errors unacceptably large. The simplest minimum basis set that is at all usable is designated Slater-type orbitals—3 Gaussians (STO-3G).

The minimum basis set may be augmented by the addition of further orbitals, the purpose of which is to correct the shapes of the orbitals to make them more realistic. Designations such as 6-31G* describe some of these *extended basis sets*. The reader may consult Hehre *et al.* (1986), Grant and Richards (1995), Leach (1996), and Levine (2013) or the review by Davidson and Feller (1986) for specific information.

The *variation principle* ensures that the more adjustable parameters the wave function contains, of course within the constraints of the problem, the lower the calculated minimum energy will be and that this minimum will not be lower than the true energy. How far one pushes the elaboration of the assumed form of the wave

(a)

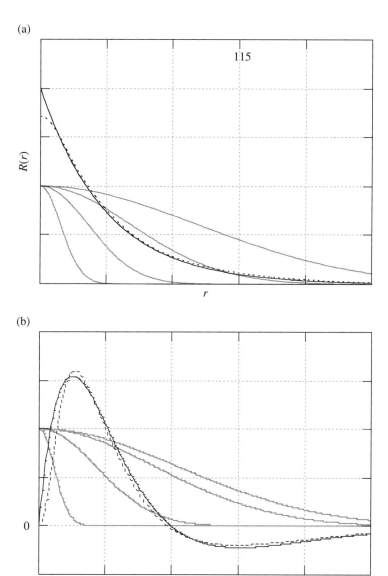

(b)

**FIGURE 5.1** Illustration of the approximation of radial parts of hydrogen-like atomic orbitals (heavy curves) as sums of Gaussians: (a) $1s$ and (b) $3p$. Not to scale and fitted by trial and error (not optimized).

function by adding other elementary functions to the minimum basis set depends on how accurate a result is needed, but it is severely limited by the number of electrons in the molecular system under investigation. The time to complete the calculation goes up rapidly as the number of electrons and the size of the basis set are increased, so rapidly that the calculation can become impossible for a molecule that in other contexts is considered of only moderate size. This remains true even though the

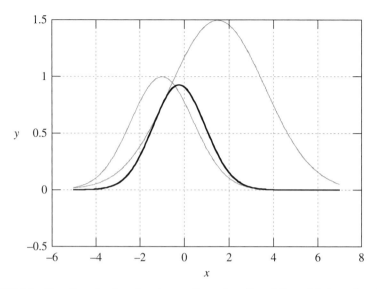

**FIGURE 5.2**   Two Gaussian functions in one dimension, $G_1$ and $G_2$, and their product (heavy curve).

symmetry of the molecule is used to eliminate duplicate calculations of integrals that must be equal and to identify any that are necessarily zero, by the methods of group theory. Similar problems obviously exist in systems containing more than a very few ordinary-sized molecules.

Increasing the size of the basis set still leaves the electrons moving in an average field and allows the calculation to approach not the true energy but the (higher) *HF limit*. Electrons avoid each other for two reasons. Firstly, the *Pauli exclusion principle* forbids two electrons of the same spin to occupy the same spatial orbital. This is ensured if the overall wave function is antisymmetric, in the sense that exchange of any two electrons changes the sign of the wave function. Equation 5.1 is a simple example:

$$\Psi = \psi_1(1)\,\psi_2(2) - \psi_1(2)\,\psi_2(1) = \begin{vmatrix} \psi_1(1) & \psi_1(2) \\ \psi_2(1) & \psi_2(2) \end{vmatrix} \tag{5.1}$$

The two orbitals $\psi_1$ and $\psi_2$ may be different space functions or identical space functions but with opposite spin. Each may be a function of the coordinates of either electron 1 or electron 2. Interchanging the electrons changes the sign only of $\Psi$. For larger systems, writing the wave function in determinant form exploits two properties of any determinant: that exchanging any two columns changes its sign and that if two columns are identical, the determinant vanishes, so that if two electrons of the same spin occupy the same region of space, the wave function will vanish. The HF method takes this constraint into account.

Secondly, Coulomb repulsion causes electrons of whatever spins to keep apart. The HF method does not allow for this second effect. There are several ways to do this, of which two are in common use. *Configuration interaction* (CI), which involves the inclusion of one or more electron configurations in which an electron is placed in an otherwise empty orbital (an excited state), accounts better for electron–electron repulsion, that is, for the fact that each electron sees the others as moving particles, rather than an average cloud of charge. Full CI is feasible only for small molecules and, even then, only when modest basis sets are used. Various restricted versions are often used. Møller and Plesset (1934) proposed a method based on perturbation theory, which does not increase in complexity for larger systems (see, e.g., Leach, 1996, pp. 83–85). Versions of the *Møller–Plesset* method are designated MP2, MP3, and so on, indicating the order of perturbation used. (First-order perturbation merely reproduces the HF energy.) Compared to CI, it has the disadvantage that it does not conform to the variational principle, that is to say the lowest energy calculated may be below the true energy, though this can be compensated for. It has the advantage of much greater speed.

If the molecule under study is polyatomic, the whole process may then be repeated with slightly different internuclear distances and angles, until a configuration of minimum total energy is found. Care must be taken to ensure that this minimum is the *global minimum*, not a local minimum that represents an unstable configuration of the molecule. If the "molecule" is an activated complex (is in a TS), the point sought is a minimum with respect to all but one of the vibrational modes. It represents a maximum in the energy along the reaction coordinate, the $x$ direction in Figure 1.4. Vibrational analysis can tell one whether a stationary point is a minimum (all force constants positive), a saddle (one negative force constant), or a point where more than one force constant is negative.

To represent a reaction, it is necessary to carry out these calculations for the reactant species and for the products. The energy change for the reaction, $\Delta U$, is obtained as a rather small difference between large numbers, so the requirements of accuracy in the calculations are severe.

Investigation of kinetic parameters of a reaction is also possible, either by postulating a reasonable structure for the TS or by exploring paths of lowest energy to find saddle points (see Section 1.7). The activation energy is obtained by difference between the energy of the saddle point and that of the initial state, with corrections for the zero-point energies, as before. As in calculation of $\Delta U$ of reaction, for $\Delta^{\ddagger}U$, high accuracy is required. Methods have been devised that combine low-level treatments of the whole molecules involved in the reaction with high-level treatments of "model systems" containing the portions of the molecules directly involved (Levine, 2013).

### 5.2.2 Semiempirical Methods

To reduce the amount of calculation required, a variety of *semiempirical* methods have been proposed. One may assume, for instance, that the core electrons (all but the valence electrons) do not change significantly in energy or in distribution about the nuclei to which they are most strongly bound, when the atoms or groups of atoms containing them are transferred from one molecule to another of related structure. It thus is possible to reduce the number of electrons for which explicit quantum-mechanical calculations are necessary, with great saving in computation time.

The atomic cores are represented by fixed charge distributions, adjusted to give correct results in simple systems. Further savings result from replacing as many integrals as possible by fixed numbers, similarly adjusted. Overlap integrals between wave functions of certain symmetries on the same atom may be taken as constant; many are zero. Overlap integrals between orbitals on adjacent atoms in a molecule are assumed to depend only on the kinds of atoms involved and their separation, and overlap integrals between nonadjacent atoms may be completely neglected. The more integrals in the ab initio scheme can be replaced with empirical parameters or set equal to zero, the greater the saving of computer time. Particular methods are labeled as, for example, *modified neglect of differential overlap* (MNDO) (Dewar and Thiel, 1977). It is common to use semiempirical methods to find a configuration of minimum energy, which may then be refined by ab initio calculation.

### 5.2.3 Density-Functional Theory

The Hohenberg and Kohn theorem (1964; Leach, 1996, pp. 528–533; Levine, 2013) states that all the properties of a molecular system in its ground state can be derived from the electron density distribution function. The total energy may be expressed as the sum of kinetic, potential, and exchange/correlation terms as in Equation 5.2, where $\rho$ is to be understood as a function of the internal coordinates, symbolized by the vector **r**:

$$E(\rho) = T(\rho) + E_{NN}(\rho) + E_{Ne}(\rho) + E_{ee}(\rho) + E_{XC}(\rho) \tag{5.2}$$

The terms are the electronic kinetic energy, the internuclear repulsive energy, the nuclear–electronic attractive energy, the interelectronic repulsive energy, and the exchange and correlation energy. The second, third, and fourth terms are easily seen to be functionals of electron density distribution only. (A function turns one number into another number: a *functional* turns a function into a number.) The Hohenberg–Kohn theorem gives assurance that calculation of the other terms is possible (but does not prescribe how). The total energy so derived is variational, that is to say it conforms to the *variational principle*: an incorrect density distribution yields an energy above the true value, provided the exact functional, which gives the relationships between $\rho$ and $T$ and between $\rho$ and $E_{XC}$ are known.

An approximation of the electron density may first be derived by a simplified ab initio calculation, in which the electrons are treated as not interacting. The electron density is obtained from the sum of the density contributions from each of the $n$ occupied spin orbitals, that is,

$$\rho(\mathbf{r}) = \sum_{i=1}^{n} \left( |\psi_i(\mathbf{r})|^2 \right) \tag{5.3}$$

Refinement of the density function, and consequently of the energy and other related properties, is then performed, using a representation of the electron density through a set of orthonormal single-electron wave functions, by the method of Kohn and Sham (1965), which takes the correlation and exchange into account.

Virtues of the density-functional method are that refinement is accomplished with much less computational effort than is required for high-level *ab initio* calculation and that the computer time required does not increase as rapidly with system size.

## 5.3  STATISTICAL-MECHANICAL METHODS

In order to calculate quantities that can be compared with experimental measurements, almost all of which relate to systems containing very large numbers of molecules, it is helpful to use statistical methods and, by the formalism of statistical thermodynamics, calculate such properties as molar enthalpies, entropies, and Gibbs energies. Two distinct methods are employed: the MC method, which can yield only equilibrium properties, and MD, which can yield information about both equilibrium properties and time-dependent processes. In these methods, a number in the range 100–10,000 of molecules is considered (far too large for QM).

The molecules are represented as objects exerting forces on their neighbors (see, e.g., Berry *et al.*, 2000, chapter 10), so that a potential energy of pairwise interaction may be defined. As was mentioned in section 5.1, the simplest form of potential energy function usable with atoms is that between hard spheres. A more realistic function, but one which is still reasonably easy to handle mathematically, is the Lennard-Jones (1924) potential, a common form of which is represented by Equation 5.4:

$$V(r) = 4\varepsilon \left[ \left( \frac{\sigma}{r} \right)^{12} - \left( \frac{\sigma}{r} \right)^{6} \right] \tag{5.4}$$

where $\varepsilon$ is the depth of the energy minimum and $\sigma$ is the collision distance, within which the potential begins to be positive. Another is the Buckingham (1938) potential, similar to the Lennard-Jones but with the repulsive $r^{-12}$ term replaced by an exponential in $r$. It is more realistic but less easy to deal with in computation. These functions are illustrated in Figure 5.3.

Modeling polyatomic molecular interactions require a more complicated function. One approach is through *interaction-site models*, which represent a molecule as an arrangement of sites or centers of force, which are usually the centers of larger atoms or groups such as methyl, each interacting with similar sites on the other molecule, as in Equation 5.5:

$$V(r,\varphi,\theta,\psi) = \sum_{i,j} 4\,\varepsilon_{i,j} \left[ \left( \frac{\sigma_{i,j}}{r_{i,j}} \right)^{12} - \left( \frac{\sigma_{i,j}}{r_{i,j}} \right)^{6} \right] \tag{5.5}$$

The index $i$ runs over the sites on one molecule, and $j$ over those on the other. The angles $\varphi$, $\theta$, and $\psi$ are the Euler angles that define the orientation of the second molecule with respect to the first.

Other versions of the multisite model use hard sphere repulsions at each site and attractive interactions of several kinds. These include the London (dispersion)

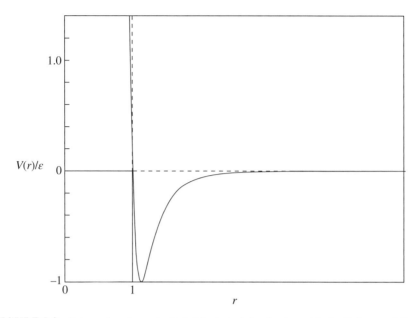

**FIGURE 5.3**   Intermolecular potentials illustrated: hard sphere (dashed), Lennard-Jones (black). A Buckingham potential adjusted to have the same collision distance, depth of well and long-range energy as the latter was indistinguishable from it at the scale of this diagram.

potential having $r^{-6}$ dependence on intermolecular separation and, if the molecule is polar, longer-range electrostatic interactions arising from charges or dipoles suitably distributed in the molecule. Figure 5.4 depicts one model for the water molecule.

More complex molecules require also that their internal motions—bending, internal rotations, etc.—be represented. The molecules are then usually treated as purely classical objects, although modern treatments may also incorporate quantum degrees of freedom in a consistent manner. The Car–Parrinello method, described as ab initio MD (Car and Parrinello, 1985), is an example.

The number of molecules should be large enough, but too large a number makes the statistical-mechanical calculation too slow. The use of periodic boundary conditions, so that a molecule that passes out of the box on one side reappears on the other, avoids the introduction of spurious boundary effects. Provided the length scales exhibited by the system are smaller than the box size, periodic boundary conditions have little effect on the predicted properties. Thus, typical fluids are readily simulated. Restricting the calculation to a "neighbor list" (only those molecules within a certain distance from a particular molecule) saves time that would be wasted on negligible interactions at long distances (see Fig. 5.5). Simulation of fluids near their critical points or of ionic liquids, in which long-range interactions are important, requires special techniques, such as Ewald summation (Ewald, 1921; Leach, 1996, pp. 294–298).

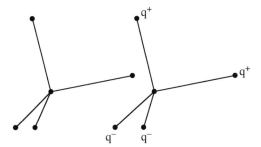

**FIGURE 5.4**  A stereo view of the ST-2 model of water (Stillinger and Rahman, 1974). Charges are placed at the tetrahedral angles: $+q$ at 100 pm from the center representing the proton positions and $-q$ at 80 pm representing the nonbonding pairs. The electrostatic potential due to these charges is in addition to a Lennard-Jones potential acting between oxygen nuclei as centers. Other models used have included further properties such as polarizability. See, e.g., Kusalik and Svishchev (1994), Svishchev *et al.* (1996).

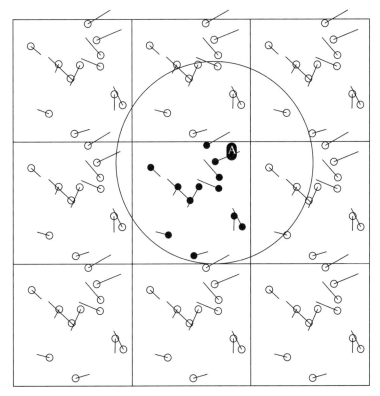

**FIGURE 5.5**  Periodic boundary conditions in two dimensions. Twelve molecules; neighbor list boundary shown for molecule **A**. The molecule entering the box just above **A** is leaving at the bottom.

### 5.3.1 MC Method

As the name suggests, the MC method is dependent on *probability*. To calculate the average value of a property $G$ (say, the energy of the system) that can be expressed as a function of all the coordinates of the molecules in a system, $G(\mathbf{r})$, it is necessary to evaluate expressions of the type

$$<G> = \frac{\int G(\mathbf{r})P(\mathbf{r})d\mathbf{r}}{\int G(\mathbf{r})d\mathbf{r}} \tag{5.6}$$

where $P(\mathbf{r})$ is the weighting (Boltzmann) function, $\exp(-U(\mathbf{r})/kT)$. To avoid wasting time on calculations for impossible or highly improbable configurations, the Metropolis *et al.* (1953) method is normally used. Starting from an arbitrary but possible configuration, small random changes in coordinates are made, governed by the probability of transition from the present configuration to one nearby. If the change in energy of the system $\Delta U$ will be negative, the new state is accepted. If the change in energy will be positive, the new configuration is accepted only if a random number $0 \le n \le 1$ is less than $\exp(-\Delta U/kT)$. Such a series of random steps is called a Markov chain. (For a discussion of the history of Markov chains, see Hayes, 2013.) After many steps, typically from $10^4$ to more than $10^6$, the mean values of calculated properties will converge to "equilibrium" values. It may be shown that simple averages over the configurations retained are properly weighted by the Boltzmann factor and give the correct results, provided that the whole of the configurational space is accessible (the process is "ergodic"). Failure to reach a portion of configuration space owing to barriers of low probability (high energy) can result in very large errors. Techniques to assure ergodicity include simply lengthening the Markov chain, increasing the number of molecules in the ensemble, starting from a number of different initial configurations, or altering the way in which random steps are made. Unfortunately, all these increase the computer time needed. Other methods can be explored, such as the "hot start," using larger perturbations at first, so the initial configuration is quickly destroyed.

Jorgensen (1982), using 125 molecules and Markov chain lengths $2.5 \times 10^5 - 7 \times 10^5$, obtained for water (at $298\,K$, $1\,atm.$) the following (experimental values in parentheses): $\Delta U_{vap}$, 41.95 (41.47) kJ $mol^{-1}$; $V_{mol}$, 18.1 (18.0) $cm^3 mol^{-1}$; $C_p$, 86.5 (75.2) J $mol^{-1} K^{-1}$; and isothermal compressibility, 35.5 (46.3) $\times 10^{-6} atm^{-1}$. Jorgensen's model of the water molecule, designated TIP4P, is a rigid "ball-and-stick" structure, with fractional charges on the **H** and **O** atoms. The repulsive and London attractive potentials are represented by a Lennard-Jones potential centered on the oxygen.

### 5.3.2 Molecular Dynamics

The method known as MD begins similarly to the MC method, by placing a number of molecules, represented by a suitable charge distribution, in a simulation cell, at time $t_0$, with assumed initial positions, orientation, and velocities. The force acting on each molecule by all the others is then calculated from the gradient of the assumed intermolecular potential. Newton's second law of motion, Equation 5.7 with finite differences replacing the differentials, is then used to calculate the acceleration of each molecule, its new velocity

$$m\frac{d^2\mathbf{r}}{dt^2} = F = -\nabla v(\mathbf{r}) \tag{5.7}$$

at time $t_0 + \Delta t$ and its new position resulting from the average velocity over the time interval $\Delta t$. The magnitude of the time step size $\Delta t$ must be chosen carefully, as too small a value will unnecessarily prolong the calculation, while too large a value will result in unphysical configurations. The calculation is repeated for a new step. After each step (or short series of steps), instantaneous values of total energy, pressure and temperature can be calculated. The process continues until the accumulated averages of these variables settle down to equilibrium values.

## 5.4 INTEGRAL EQUATION THEORIES

It can be shown (see, e.g., Hansen and McDonald, 1986, section 2.5) that all thermodynamic quantities can be calculated from the radial distribution function, $g(\mathbf{r})$. This function, which can be obtained experimentally from X-ray or neutron diffraction, can also be calculated, provided the intermolecular potential function is known. The aim of the integral equation theories is to calculate the radial distribution function from the intermolecular potential $V(\mathbf{r})$. There are several versions that differ according to the choice of simplifying assumptions made. They each result in an equation that contains $g(\mathbf{r})$ associated with the Boltzmann factor $\exp(-\beta v(\mathbf{r}))$. For example, the PY equation (Hansen and McDonald, 1986, p. 119; Percus and Yevick, 1958)

$$\exp[-\beta V(\mathbf{r})]\,g(\mathbf{r}) = 1 + \rho\int[g(\mathbf{r}-\mathbf{r}')-1](1-\exp[-\beta V(\mathbf{r}')])\,g(\mathbf{r}')\,d\mathbf{r}' \tag{5.8}$$

Here, $\beta$ is $1/kT$. The hypernetted chain (HNC) equation (van Leeuwen *et al.*, 1959) and the equation of Born and Green (1949) involve similar terms, though they are in logarithmic form. The Born–Green equation also contains the first derivative with respect to $\mathbf{r}$, so it is an integrodifferential equation for $g(\mathbf{r})$ in terms of $V(\mathbf{r})$.

## 5.5 SOLVATION CALCULATIONS

Going from pure liquids to solutions, by either MC or MD methods, the pure solvent is first simulated, and then one of the solvent molecules in the model is replaced by a solute molecule. After a number of cycles to allow the system to relax to accommodate the intruder, the chain of calculations continues until convergence of the average properties is again achieved. The solvation energy is the difference between the energies of the systems with and without the solute molecule. It is a small difference between two large numbers, so subject to large relative error. Nevertheless, results have been obtained. Simkin and Sheikhet (1995) surveyed results of calculations of solvation of a number of hydrocarbons, polar nonelectrolytes, and electrolytes in water and a few other solvents. The degree of agreement with experimental values is highly variable. Reasons for the sometimes very large errors are discussed.

Recently, both ab initio and semiempirical quantum-mechanical treatments have been adapted so as to take into account the role of solvent. Fundamentally, the SE would have to be modified so as to bring the isolated molecule from its state in vacuo to the solution state, that is, to implement the perturbation of the wave function of the molecule by the solvent. Various methods for achieving this goal have been proposed. Representation of the solvent as a continuous dielectric, with a solute contained in a cavity, was pioneered by Born (1920) for ions and by Kirkwood and Onsager for dipolar molecules (Onsager, 1936) and implemented as the quantum-Onsager self-consistent reaction field (SCRF) method. The electrostatic contribution to the energy for a dipole $\mu$ in a spherical cavity of radius $a$ immersed in a medium of relative permittivity $\varepsilon_r$ is given by Equation 5.9:

$$\Delta U_{elect} = -\frac{(\varepsilon_r - 1)}{8\pi\varepsilon_0(\varepsilon_r + 1/2)a^3}\mu^2 \tag{5.9}$$

It arises from the *reaction field*, that is, the field within the cavity due to the charges induced in the wall of the cavity by the dipole, acting back on the dipole.

If the Hamiltonian of the molecule in vacuo is $H_0$, the Hamiltonian of the same molecule in the cavity may be written: $H_0 + H_{rf}$, where $H_{rf}$ is given by Equation 5.10 (Tapia and Goscinski, 1975):

$$H_{rf} = -\mu^T \frac{(\varepsilon_r - 1)}{(\varepsilon_r + 1/2)}\langle \Psi | \mu | \Psi \rangle \tag{5.10}$$

Here, $\mu$ and $\mu^T$ are the matrix representation of the dipole moment operator and its transpose. This calculation can be implemented by available programs, such as "Gaussian-92," described by Foresman and Frisch (1993), or "Jaguar," described by Bochevarov *et al.* (2013). Foresman and Frisch give as an example (p. 193) the calculation of $\Delta U$ for the *gauche–anti* rotation of 1,2-dichloroethane in the gas phase and in solution in cyclohexane ($\varepsilon_r = 2$), using the 6-31 + $G(d)$ basis at both the HF and MP2 levels. Their results are summarized in Table 5.1.

Wong *et al.* (1992) used this approach to calculate the shifts of the tautomeric equilibrium between 2-pyridone (**1**) and 2-hydroxypyridine (**2**) in passing from the gas phase to cyclohexane ($\varepsilon_r = 2.0$) and to acetonitrile (35.9). The calculated (and experimental) values for the enol/keto ratio were as follows: gas, 2.9 (3.9); cyclohexane, 0.54 (0.57); and acetonitrile, 0.020 (0.007).

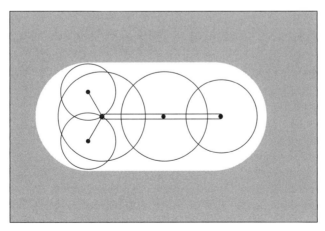

**TABLE 5.1   Calculated (Hartree–Fock, Second-order Møller–Plesset) and Experimental Internal Energy of _Gauche–anti_ Rotation of 1,2-dichloroethane in Vacuo and By QO-SCRF in Solution[a]**

| Medium | $\Delta U_{rot}$ kJ mol$^{-1}$ | | |
|---|---|---|---|
| | HF | MP2 | Exp't |
| Gas ($\varepsilon_r = 1$) | 8.2 | 6.32 | 5.02 |
| Cyclohexane ($\varepsilon_r = 2$) | 5.52 | 4.14 | 3.81 |

[a] From Foresman and Frisch (1993).

Refinements of the quantum-Onsager SCRF method add the higher electric moments—quadrupole, octupole, etc.—and use an ellipsoidal rather than a spherical cavity. For molecules that are not even approximately spherical, a cavity following the van der Waals envelope of the molecule (similar to that illustrated in Fig. 5.6) may be used (Miertus _et al._, 1981) at the expense of much more complicated calculation. Another complication is pointed out by Muddana _et al._ (2013), in that ionic or neutral polar solutes may cause nonlinear responses. In explicit simulations of water containing large globular molecules with exposed polar groups such as OH, electrostriction resulted

**FIGURE 5.6**   Cavity in a continuous dielectric (schematic) containing a ketene molecule. Rotation about the molecular axis is not hindered.

in a stronger linear reaction field, which they contrast with the weaker-than-linear response due to dielectric saturation expected with small ionic solutes.

Hofer *et al.* (2008) describe the quantum-mechanical charge field (QMCF) method for solvation calculations. They distinguish a core region containing the solute molecule or ion and the nearest solvent molecules treated by QM, an outer region treated by MD (discussed in section 5.3.2), and an intermediate layer of solvent in which QM is used to calculate local charge distributions at each step of the MD calculation to effect coupling between the inner and outer regions. Examples of results on hydration of variously charged species were reviewed. For instance, it was shown that supposedly square-planar coordinated Pt(II) and Pd(II) cations in water have additional, more labile water molecules in the axial positions, that is, they are in a distorted octahedral environment. Hydration of anions presented some difficulties, owing to weaker coulombic attraction and H bonding. Low-level (HF) QM predicted too long and too weak H bonds; on the other hand, density-functional methods predicted H bonds that were too rigid. Among neutral molecules, the calculated aqueous solubility of $CO_2$ was approximately correct. Weak and few H bonds were found, about 0.6 per O of $CO_2$.

## 5.6 SOME RESULTS

The whole field of theoretical calculation as it relates to solutions and solvent effects is currently very active. A useful review is that by Rotzinger (2005). It is not possible in a brief compass to mention all the important advances. We present a random selection of examples of this work, with no particular theme and without any attempt to be comprehensive.

### 5.6.1 Microsolvation

By *microsolvation*, we mean the association in the gas phase of a single central molecule or ion with a small number of another species that may be considered a ligand or a potential solvent. For calculation of solvation energies quantum mechanically, one would like to be able to start with a "solute" molecule, **A**, and a number of "solvent" molecules, **B**, and calculate to a certain level of theory the energy of **A** and **B** alone and then of a series of *supermolecules* or clusters **A.B**, **A.2B**, **A.3B**, and so on and a series of clusters 2**B**, 3**B**, and so on until the energy difference represented by

$$\mathbf{A} + n\mathbf{B} \rightarrow \mathbf{A}.n\mathbf{B}; \Delta U_n$$

becomes independent of $n$. This difference, then, should be the solvation energy of a molecule of **A** in the solvent **B**. In principle (see preceding text), one should be able to calculate this energy to any desired degree of accuracy. Unfortunately, to do this by pure quantum-mechanical methods would strain the resources of any presently available computer, as the number of molecules needed to reach the constant difference would be large. By severely limiting the number of solvent molecules, results can be obtained, however, that may be compared with experimental data on reactions in the gas phase by mass-spectrometric methods.

**TABLE 5.2   Enthalpy Changes for Gas-phase Reactions:** $M^+(H_2O)_{n-1} + H_2O = M^+(H_2O)_n$ **and** $X^-(H_2O)_{n-1} + H_2O = X^-(H_2O)_n$ [a]

| | | | | | | $-\Delta H_{n-1,n}$ (kJ mol$^{-1}$) | | | | | |
|---|---|---|---|---|---|---|---|---|---|---|---|
| $n$ | H$^+$ | Li$^+$ | Na$^+$ | K$^+$ | Rb$^+$ | Cs$^+$ | OH$^-$ | F$^-$ | Cl$^-$ | Br$^-$ | I$^-$ |
| 1 | 690 | 142 | 100 | 75 | 67 | 57 | 94 | 97 | 55 | 53 | 43 |
| 2 | 151 | 108 | 83 | 67 | 57 | 52 | 69 | 69 | 53 | 51 | 41 |
| 3 | 93 | 87 | 66 | 55 | 51 | 47 | 63 | 57 | 49 | 48 | 39 |
| 4 | 71 | 69 | 58 | 49 | 47 | 44 | 59 | 56 | 46 | 46 | |
| 5 | 64 | 58 | 51 | 45 | 44 | | 59 | 55 | | | |
| 6 | 54 | 51 | 45 | 42 | | | | | | | |
| 7 | 49 | | | | | | | | | | |
| 8 | 43 | | | | | | | | | | |

[a] Kebarle (1972). Converted to kJ mol$^{-1}$.

Mass-spectrometric studies by Kebarle and coworkers (Arshadi *et al.*, 1970; Kebarle, 1972) have enabled determination of the enthalpies of successive reactions (in the gas phase at low pressures, in the presence of a gas such as nitrogen or methane, which functions both as a heat bath and as a chemical ionization carrier) of the types

$$M^+(H_2O)_{n-1}(g) + H_2O(g) = M^+(H_2O)_n(g)$$

and

$$X^-(H_2O)_{n-1}(g) + H_2O(g) = X^-(H_2O)_n(g)$$

where M$^+$ is H$^+$ or an alkali-metal ion and X$^-$ is a halide ion. Their results are summarized in Table 5.2. Later work from the same laboratory (Lau *et al.*, 1982) refined the values for H$^+$.

In Figure 5.7, the stepwise enthalpy changes for H$^+$ are plotted against the reciprocal of the number of water molecules after the first in the product ion, $1/(n-1)$. With the exception of the point for $n=4$, which is above the line, suggesting unusual stability for the ion $H_9O_4^+$ (Berry *et al.*, 2000, p. 309), the values decline regularly, tending toward a value somewhat below the enthalpy of vaporization of bulk water. (This is presumably because the molar enthalpy of vaporization from a very small droplet is less than the bulk value.) This experimental system is fully analogous to the quantum-mechanical model used by Newton (1997). His ab initio (4-31G) calculated results for $\Delta U$ for these reactions are also shown in Figure 5.7. They also show a marked drop in the magnitude of the energy change for the addition of the fourth water molecule.

The similar hydration of hydroxide ion greatly reduces its energy, an observation that is consonant with the great increase in the chemical activity of OH$^-$ as the concentration of water in a solution of NaOH in mixed DMSO/water solvent is decreased (see Chapter 6).

The successive values of $-\Delta H_{n-1,n}$ for the alkali hydrates recorded in Table 5.2 fall off smoothly, showing no favored coordination number (Fig. 5.8). Notably, this is

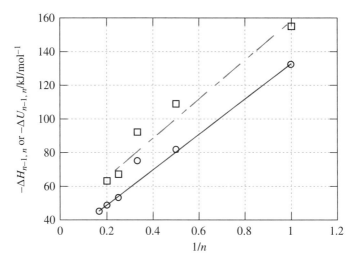

**FIGURE 5.7** Stepwise experimental, gas-phase enthalpies of hydration of $H_3O^+$, circles (Lau *et al.*, 1982), and corresponding theoretical energies, squares (Newton 1997), plotted versus the reciprocal of the number of water molecules. NB, $n$ in this and the following figure corresponds to $n-1$ in Table 5.2.

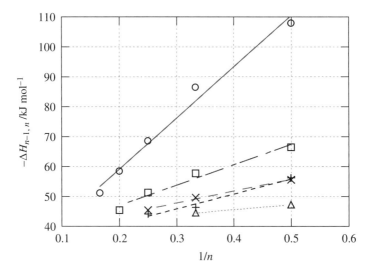

**FIGURE 5.8** Stepwise experimental, gas-phase enthalpies of hydration of alkali-metal ions (Kebarle, 1972) plotted versus the reciprocal of the number of water molecules. Circles, $Li^+$; squares, $Na^+$; oblique crosses, $K^+$; plus signs, $Rb^+$; triangles, $Cs^+$.

true for $Li^+$ out to $n=6$ (one might have expected $CN=4$ to be favored). The halide ion hydrates show a similar but less steep descent from lower initial values. Fluoride ion is an exception, as the first two water molecules form particularly strong hydrogen bonds with this ion.

Kebarle cites calculations using a model involving Lennard-Jones (6, 12) potential energy terms, plus ion–dipole, ion–induced-dipole, and dipole–dipole terms that agreed well with most of the experimental findings for the alkali-metal ion hydrates. It was found necessary to allow explicitly for some covalent interaction between sodium or (especially) lithium and the first two or three water molecules.

### 5.6.2 A Classic Reaction: The $S_N2$ Reaction, Gas-Phase versus Solution

Probably the most intensively studied reaction, both experimentally and through calculations, is the $S_N2$ nucleophilic substitution of the general type

$$Y^- + R - X \rightarrow Y - R + X^-$$

These reactions show great sensitivity to the nature of the solvent medium in which they are conducted, with respect to reaction rates, equilibria, stereochemical outcome, and even the formation of side products, for example, via elimination. All these different aspects have been subject to theoretical investigations. The reader is referred to the excellent text by Shaik *et al.* (1992) for a full discussion of the different theoretical models that have been used and the derived results.

Of special interest are the energy barriers for given reactions, and in principle, calculation of the potential energy surfaces can provide these quantities. In practice, the majority of studies report one-point calculations on SCF-minimized geometries. Generally, calculations for gas-phase $S_N2$ reactions use the model depicted for the identity reaction:

$$X^- + H_3C - {}^*X \rightarrow [X \text{---} CH_3 \text{---} X]^- \rightarrow X - CH_3 + {}^*X^-$$

The reaction profile for reactants (**R**) going to products (**P**), shown schematically in Figure 5.9, shows a double-well potential, in which the minima $\mathbf{C}_1$ and $\mathbf{C}_2$

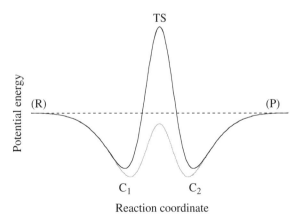

**FIGURE 5.9** Reaction profile (schematic) for an $S_N2$ reaction in the gas phase, showing potential wells corresponding to ion–dipole complexes. The energy of the transition state may by higher (heavy curve) or lower (light curve) than the energy of the reactants. Entropy effects are important. See discussion in Shaik *et al.* (1992) or Olmstead and Brauman (1977).

represent *ion–dipole complexes*, separated by a central energy barrier representing the transition state (TS) containing a pentacoordinated carbon atom. Computations of the central barrier height above the energy of the well-separated reactants have been performed for a number of gas-phase identity reactions, using ab initio as well as various semiempirical methods (see Table 5.3) (based on Shaik *et al.*, 1992, table 5.2, p. 165). The results obtained by various authors show the difficulty of obtaining precise agreement between methods in such calculations, which involve differences between large numbers.

**TABLE 5.3   Identity S$_N$2 Reactions: Comparison of Central Barrier Heights Calculated By Various Methods[a]**

| | | Ab Initio | | | Semiempirical | | |
|---|---|---|---|---|---|---|---|
| Entry | X | 3-21G | 4-31G | Large basis sets and CI | MNDO | AM1 | INDO($\lambda$) |
| 1 | F | 51 | 49 | 71,[b] 75–84,[c] ≈88,[d] ≈71,[e] 71.5,[f] 84,[g] 54[h] | 188 | — | 91.6 |
| 2 | Cl | 21 | 23.0 | 58.2[1], ≈50[4], ≈66[5], 58,[i] 53,[j] 26,[k] 72[8] | 44 | 37.9 | 91.6 |
| 3 | Br | | | 53.1[7], 48.5[6] | 33.5 | 23 | |
| 4 | I | | | | | 10.0 | |
| 5 | HOO | | 77 | | | | |
| 6 | FO | | 79 | | | | |
| 7 | HO | | 89 | | 201 | | 1.7 |
| 8 | CH$_3$O | | 98 | | 223 | | |
| 9 | HS | | 65 | | | | |
| 10 | HCC | | 210 | | | | |
| 11 | NC | | 183 | | 113 | | |
| 12 | CN | | 119 | | | | |
| 13 | NH$_2$ | | ≈160 | | | | |
| 14 | H | | 218 | 255–278[2], 230[l] | −8 | | 216 |
| 15 | NH$_3$[m] | | 78 | | | | |
| 16 | H$_2$O[13] | | | 42[n] | | | |

Values are converted to kJ mol−1.

[a] From Shaik *et al.* (1992), table 5.2, p. 165. See references therein.

[b] 6-31G*//6-31G*.

[c] Triple zeta + diffuse functions.

[d] Double zeta + diffuse functions.

[e] Double zeta + diffuse functions + MP4/SDTQCI corrections.

[f] Double zeta = diffuse functions + polarization functions + MRD/CI.

[g] Double zeta = diffuse functions + polarization functions.

[h] MP2/6-31+G* optimization.

[i] 6-31 + G* + MP4//4-31G.

[j] 6-31G* + MP4//4-31G.

[k] 4-31G + MP4//4-31G.

[l] 6-31 ++G** + MP2.

[m] For the neutral nucleophiles NH$_3$ and H$_2$O, the reaction is as follows: X: + CH$_3$–X = X–CH$_3$ + X:

[n] 6-31G** + MP3//3-21G including ZPE correction.

The major energetic change when one goes from the gas phase to a solution phase, especially in aqueous or alcoholic solvents, is stabilization of the ionic species $X^-$ through H bonding. The charge-dispersed TS becomes stabilized to a much lesser extent. If the condensed-phase reaction is performed in an aprotic solvent that is, however, highly polar, such as DMF or DMSO, then the ions $X^-$ can no longer be solvated through H bonding and will be much less stabilized. This is illustrated in Figure 5.10, which depicts the results of calculation for the identity reaction

$$Cl^- + H_3C - {}^*Cl \rightarrow Cl - CH_3 + {}^*Cl^-$$

in the gas phase, in water, and in DMF.

"Real" $S_N2$ reactions of interest are, of course, nonidentity reactions. Various investigations have reported on central energy barriers and geometries of the transition structures, as well as reaction energies, both for the gas-phase and aqueous

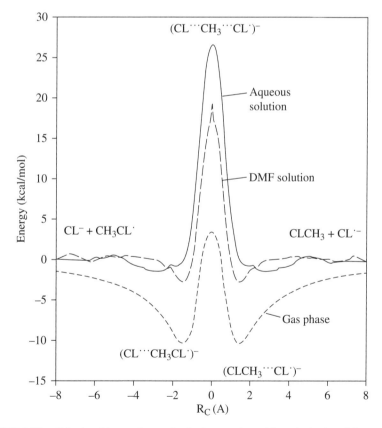

**FIGURE 5.10**  Calculated internal energies in the gas phase (short dashes) and the potential of mean force in DMF (long dashes) and in aqueous solution (solid curve) for the reaction of $Cl^-$ with $CH_3Cl$ as a function of the reaction coordinate, $r_c$, in angstroms. Source: Chandrasekhar and Jorgensen (1985) with permission of the American Chemical Society.

reactions. Again, the reader is referred to the text by Shaik *et al.* (1992). The energetics of the reaction of methyl bromide with hydroxide ion microsolvated by 0, 1, 2, 3, etc. water molecules have been investigated by Bohme and Mackay (1981) using mass spectrometry. (See Reichardt's discussion of the energetics of this and related reactions (Reichardt, 1988, pp. 133–136; 2003, pp. 156–162.))

### 5.6.3   Solvatochromism: Theoretical Calculations

Several authors have carried out theoretical calculations of the solvation of molecules that form the bases of solvatochromic scales of solvent properties. To take one example, we may consider the molecule 4-[(4′-hydroxyphenyl)azo]-*N*-methylpyridine (**3**) (called "Buncel's dye" by Rauhut *et al.*, 1993), which exists as a resonance hybrid of two structures.

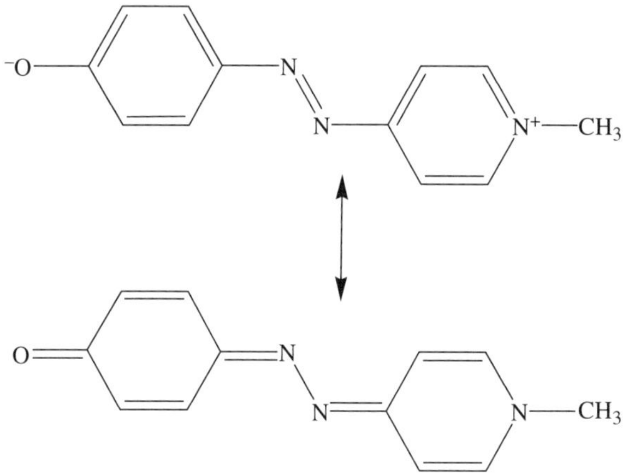

**3**

A UV-visible spectroscopic study of **3** and related substances revealed a strong solvatochromic effect, which served as the basis of the establishment of a $\pi^*_{azo}$ solvent polarity scale (Buncel and Rajagopal, 1989, 1990, 1991). The theoretical study of Rauhut *et al.* (1993) was based on AM1 methodology (Dewar and Storch, 1985, 1989) but used a double electrostatic reaction field in a cavity, dependent on both the relative permittivity and the refractive index. Nuclear motions interact with the medium through the relative permittivity, but electronic motions are too fast; only the extreme high-frequency part of the dielectric constant is relevant. These authors were able to evaluate solvent-specific dispersion contributions to the solvation energy. The calculations reproduced satisfactorily the experimental solvatochromic results for **3** in 29 different solvents. The method has also been successfully applied to other solvatochromic dyes, including Reichardt's $E_T(30)$ betaine.

Matyushov *et al.* (1997) calculated the energy changes that form the bases of the $\pi^*$ and $E_T(30)$ scales, using a model in which the solvent molecules were represented as hard spheres and the solute (dye) molecules as hard hemisphere-capped

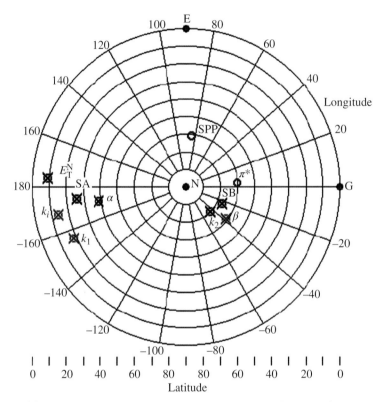

**FIGURE 4.3** Hemisphere projection: three rate constants and seven solvent parameters. Circles represent points in their natural positions, crossed circles antipodes of points that fall in the hidden hemisphere, that is, the negatives of the indicated variables.

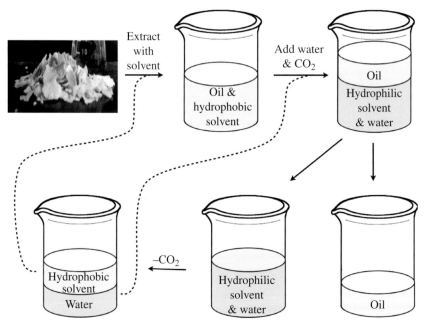

**FIGURE 8.1** The process by which a switchable-hydrophilicity solvent can be used to extract soybean oil from soybean flakes without a distillation step. The dashed lines indicate the recycling of the solvent and the aqueous phase. Source: Jessop *et al.* (2010) by permission of the Royal Society of Chemistry.

cylinders, and only relatively long-range solute/solvent interactions were considered. Interactions considered were permanent dipole–dipole (*perm*), London dispersion (*disp*), and dipole–induced dipole (*ind*). The polarizabilities were assumed to be isotropic. Both $\pi^*$ and $E_T(30)$ were shown to contain contributions from all three interactions. The interesting result emerged that while both scales are similarly affected by *disp* and *perm*, the *ind* term reinforces the other two in the case of $\pi^*$ but acts in the contrary sense with $E_T(30)$. The reason adduced is that in passing from the ground to the excited state, the dipole moment and the polarizability of 4-nitroanisole (taken as the type dyc for $\pi^*$) both increase, whereas with the betaine-(30) the polarizability increases considerably but the dipole moment decreases. The agreement between the experimental and calculated values of both parameters is good, except that $\pi^*$ is overestimated for the more polarizable chlorinated and aromatic solvents, while $E_T(30)$ is underestimated for the alcohols. The latter discrepancy is undoubtedly due to the explicit neglect of hydrogen bonding and is in keeping with the observation that $E_T(30)$ falls between the dipolar and acidic groups of parameters in Figure 4.2.

The position of $E_T^N$ (or its nonnormalized version $E_T(30)$) between the acidic and polar groups, mentioned in the previous chapter, has been explicitly studied by Laurence and coworkers (Cerón-Carrasco *et al.*, 2014a, b). Values of $E_T(30)$, the first electronic transition of the dye 30, were calculated by time-dependent density-functional theory, using a model in which the dye is in a cavity in a solvent represented as a polarizable, continuous dielectric. For non-HBD solvents, a linear regression between the calculated and observed values of $E_T(30)$ was obtained. For every HBD solvent, the observed value lay above the regression line by an amount that was taken to be a measure of the HBD acidity of the solvent. The symbol $\alpha_1$ represents this measure, expressed in normalized (dimensionless) form by dividing all by 12.87 kcal/mol, setting the value for methanol at 1.00. Values are tabulated for 55 solvents, including examples of C–H, N–H, and O–H donors. A comparison with $\alpha$ (Taft–Kamlet) is shown in Figure 5.11. Alcohols, carboxylic acids, and water (O–H acids) lie about a straight line. The others, C–H and N–H acids, lie close to another line, below the first. The slopes are the same within the uncertainties: lower, $0.69 \pm 0.02$, and upper, $0.8 \pm 0.1$. Using the revised set of values of $\alpha$ reported by Marcus (1993), however, Laurence has pointed out this dichotomy is not sustained (C. Laurence, 2014, private communication).

### 5.6.4 Solvation of Organic Molecules

Among the applications of integral equation theory to solutions, Koga *et al.* (1996) used a reference interaction-site model (RISM), an approximation due to Chandler and Andersen (1972), to calculate the local solvation behavior of naphthalene in supercritical carbon dioxide. They used a ten-site model for naphthalene and a three-site model for carbon dioxide, in which partial charges were placed on the sites in addition to Lennard-Jones potentials acting between sites on different molecules. Their potential function was a sum over the ten naphthalene sites and the three carbon dioxide sites of terms of the form of Equation 5.11:

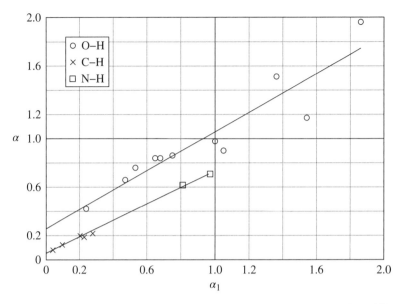

**FIGURE 5.11**    Taft–Kamlet $\alpha$ versus $\alpha_1$, the H-bond acidity scale based on the acidity contribution to the solvatochromism of Reichardt's dye 30 (Cerón-Carrasco *et al.*, 2014a, b). Acids of types O–H are shown as circles, C–H acids as crosses, and N–H as squares.

$$v_{i,j} = 4\varepsilon_{i,j} \left[ \left( \frac{\sigma_{i,j}}{r_{i,j}} \right)^{12} - \left( \frac{\sigma_{i,j}}{r_{i,j}} \right)^{6} \right] + \frac{z_i z_j e^2}{r_{i,j}} \tag{5.11}$$

They demonstrated that the RISM integral equation and MC methods gave comparable results and that the nearest $CO_2$ molecules showed preferred orientations in certain positions: parallel to the naphthalene molecular plane and to the long axis when above the plane but perpendicular to the plane whether "abeam" or "ahead" of the naphthalene. They also obtained the partial molal volume of naphthalene as a function of the density of solvent; it is negative at low densities, owing to a number of $CO_2$ molecules being attracted to naphthalene, increasing to reasonable positive values as density increases.

In a study of aqueous solvation of *cis-* and *trans-N*-methylacetamide, Yu *et al.* (1991) compared MC, MD, and several integral equation methods with each other and with experimental data. It was apparent that the various integral equation methods differed in their applicability to this system. The HNC method in particular, though recommended by some for use with polar substances, gave poor results for the absolute value of the solvation free energy for both conformers (the wrong sign, owing to overestimation of the entropy contribution), though the difference between them was close to the experimental value.

### 5.6.5 Bimolecular Reactions: Correcting for Suppressed Translational and Rotational Entropies

Bimolecular reactions have large entropic effects in the gas phase due to large changes in the translational and rotational degrees of freedom that occur during the course of a reaction. In solution, translational and rotational degrees of freedom are highly suppressed owing to the interactions with solvent molecules. Such explicit solvent/solute interactions are not captured with continuum solvation models. Consequently, thermodynamic corrections to potential energies obtained through frequency calculations performed using continuum solvation models often overestimate the contributions of translational and rotational degrees of freedom to the entropy. A method to approximate the free energy for solution-phase bimolecular reactions has been proposed by Sakaki's group (Sumimoto *et al.*, 2004; Tamura *et al.*, 2003), wherein only vibrational contributions to entropy are considered, with the exaggerated translational and rotational contributions discarded. It is argued that the resulting $\Delta G_{\text{calc, corrected}}$, while an underestimate of the true value, is closer to the experimentally derived $\Delta G$ than the uncorrected calculated free energy.

This correction approach has been used in research work outlining the mechanism of a palladacycle-catalyzed methanolysis of methyl parathion (**4**) (Liu *et al.*, 2010).

**4**

The initial step of the mechanism involves a bimolecular ligand exchange process, which is rate limiting. This is followed by a second step in which attack on the phosphorous atom by a methoxide leads to expulsion of the aryl leaving group. The free energies of the initial step were calculated using continuum solvation models to describe the system in methanol, and the associated changes in free energy are given in Figure 5.12. When all contributions to the free energy are included, the calculated free energy barrier of $\Delta G_{\text{calc, uncorrected}} = 28.3\,\text{kcal mol}^{-1}$. This value is inconsistent with the experimentally determined barrier of $12.9\,\text{kcal mol}^{-1}$. Adjusting the free energies of the ground state (labeled GS in the figure) and the TS leads to a free energy barrier of $\Delta G_{\text{calc, corrected}} = 10.6\,\text{kcal mol}^{-1}$. This value is in much better agreement with the experimental value than $\Delta G_{\text{calc, uncorrected}}$. In general, it is expected that the experimental barrier will lie between $\Delta G_{\text{calc, corrected}}$ and $\Delta G_{\text{calc, uncorrected}}$, with the former providing a better estimate of the experimental value. Of course, the ability to obtain good agreement with experimental values also depends on selecting proper and appropriate methodology for the electronic structure calculation.

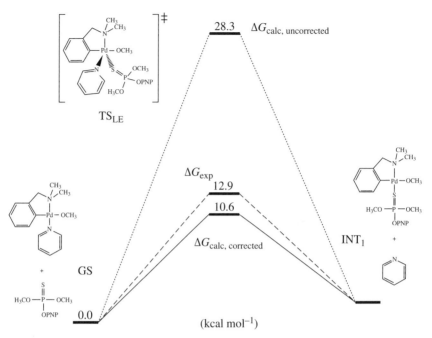

**FIGURE 5.12** Gibbs energy reaction profile of the palladacycle-catalyzed methanolysis of methyl parathion (**4**) showing the Gibbs energy of activation calculated with and without inclusion of the rotational and translational contributions to the entropy change and the experimental value. Source: Guillot *et al.* (1991) with permission of the American Institute of Physics.

### 5.6.6 Hydrophobic Solvation

The hydrophobic effect, mentioned in Section 2.7, has received much attention. Némethy and Scheraga (1962) were able to demonstrate that the decrease in entropy accompanying the entry of a hydrocarbon into aqueous solution was due to increased hydrogen bonding among the water molecules in the layer nearest the solute molecule. Okazaki *et al.* (1979), whose work was mentioned in Chapter 2, obtained for the transfer functions of methane from the gas phase to water at 25°C the following: $\delta A = 0.8$ (almost identical with the experimental value based on the solubility $2.48 \times 10^{-5}$ mol fraction), $\delta U = -1.7$, and $T\delta S = -2.5$ (all in kJ mol$^{-1}$), showing that, while there is a decrease in energy from the increased H bonding, the loss of entropy dominates the solubility. They later found (Okazaki *et al.*, 1981) the trend in stabilization energy to be not monotonic as the size of the solute increases: ethane was found to be less stabilized than methane, but pentane more stabilized, and structural details change. Abraham (1982) found, in the course of a statistical analysis of the experimental solubilities and their temperature dependence of a large number of nonpolar gases in many solvents, that for normal hydrocarbons the hydration of the terminal methyl groups is entropy dominated but that for the methylene groups is enthalpy dominated. Tanaka (1987), as a result of a study by MC and by the RISM integral equation method, reported that for nonpolar solutes of increasing size, H bonding increased, with resulting increase of solubility.

Other workers have used various methods (MC, MD, integral equations) to investigate thermodynamic aspects of hydrophobic solvation (Abraham, 1982; Durrell and Wallqvist, 1996; Guillot *et al.*, 1991). Some have examined effects of solute size or curvature (Ashbaugh and Paulaitis, 1996; Chau *et al.*, 1996; Okazaki *et al.*, 1981; Tanaka, 1987). Others have focused on specific aspects such as heat capacity (Madan and Sharp, 1996), volume, and compressibility (Matubayasi and Levy, 1996).

### 5.6.7  Solvation of Small Molecules

For a final example, Guillot *et al.* (1991), in a series of MD simulations of solutions of the rare gases and methane in water and in methanol, showed that the solvent molecules in the first shell are tangentially oriented (as depicted in Fig. 2.4b). Figure 5.13 shows the pair distribution functions for argon in water: $g_{AO}(r)$ for the distribution of oxygen and $g_{AH}(r)$ for the distribution of hydrogen, near argon. The first maxima in the two distributions are nearly coincident. Meng and Kollman (1996) presented Figure 5.14, which shows the same phenomenon in the distribution of O and H around methane in water. Guillot's group (1993) also showed that the order of solubilities—Ne<Ar, etc.—is dominated by $\Delta U$ and that $\Delta S$ values are temperature dependent in a manner that gives rise to minima in some of the solubilities as functions of temperature.

Bagno (1998) presented a diagram (reproduced as Fig. 5.15) showing clearly that in all solvents examined, the enthalpy and entropy of solvation of methane are both negative and that at ordinary temperatures the $T\Delta S$ term is substantially larger than the $\Delta H$ term (so the solubility may be described as "entropy dominated"). In the

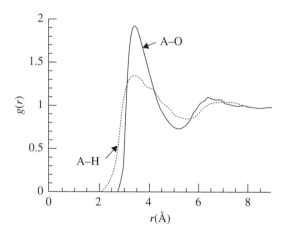

**FIGURE 5.13**   Calculated pair distribution functions for argon in water: argon–oxygen $g_{Ar-O}$ and argon–hydrogen $g_{Ar-H}$ (Guillot *et al.*, 1991). Note the near coincidence of the first maxima. Source: Meng and Kollman (1996) Reprinted with permission of the American Chemical Society.

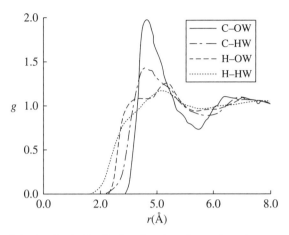

**FIGURE 5.14**   Calculated pair distribution functions for methane in water: carbon–oxygen $g_{C-O}$, carbon–hydrogen(water) $g_{C-Hw}$, hydrogen(methane)–hydrogen(water) $g_{Hm-Hw}$, and hydrogen(methane)–oxygen $g_{Hm-O}$. Note the near coincidence of the first maxima for C–Hw and C–O. Reproduced from Meng and Kollman (1996) with permission of the American Chemical Society.

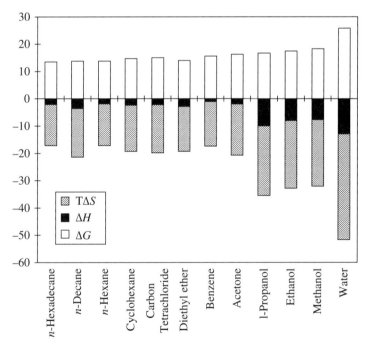

**FIGURE 5.15**   Enthalpy and entropy contributions to the Gibbs energy of solvation of methane in a number of solvents (Bagno, 1998). Reproduced with permission from the author. Courtesy of A. Bagno.

hydroxylic solvents, and especially in water, both terms are large, the entropy remaining dominant.

It appears from these studies that the picture of an "iceberg" around the solute in hydroxylic solvents is overdrawn. The shell of tangentially oriented molecules is only a monolayer, and though these molecules are restricted in their radial motion (giving rise to the entropy loss), they are not strongly ordered within the monolayer.

## PROBLEMS

**5.1** Prove that the product of any two Gaussians in three dimensions is also a Gaussian. (Since a coordinate system is arbitrary, it will suffice to set up the original two on the $x$-axis, at suitable distances on either side of the origin. The result is particularly simple, and no generality is lost, if they are centered at $-x_1$ and $+x_2$, with $x_1/x_2 = s_1/s_2$, where s represents the standard deviation.)

# 6

# DIPOLAR APROTIC SOLVENTS

## 6.1 INTRODUCTION

Dipolar aprotic solvents came into prominence in the 1960s, with the discovery that the rates of a number of reactions of the $S_N2$ type, and some other reactions as well, were very greatly enhanced, by factors up to $10^9$, when carried out in solvents such as dimethylformamide and dimethyl sulfoxide as compared to common hydroxylic solvents such as water or ethanol. This was the pioneering work of Parker (1962, 1967, 1969), Ritchie (1969), Cram (1965), and others (Coetzee and Ritchie, 1969). A list of the more common dipolar aprotic solvents together with some of their important physical characteristics is given in Table 6.1.

It is immediately obvious on glancing at the entries in Table 6.1 (all organic molecules) that all these solvents actually contain hydrogen atoms, albeit these are bound to carbon atoms, so the term "aprotic" is actually a misnomer. In principle, virtually all C–H–containing compounds can be deprotonated in the presence of a sufficiently strong base, so the solvents listed in Table 6.1 can be termed carbon acids, though mostly very weak carbon acids. Importantly, unlike the typical "normal" acids in which the ionizable hydrogen is bound to the electronegative atom oxygen, and which are therefore strong hydrogen-bond donors (i.e., HBD solvents), the C–H acids are extremely poor hydrogen-bond donors. Therefore, an alternative terminology for the dipolar aprotic solvents that has been proposed is *dipolar non-HBD* solvents (Mashima *et al.*, 1984). However, in the present account the more common naming will generally be retained.

As shown in Table 6.1, the distinguishing features of the dipolar aprotic solvents are the relatively high relative permittivities ($\epsilon > 15$), dipole moments ($\mu > 5$), and $E_T^N$ solvent parameters ($E_T^N > 0.3$).

Clearly, anions will be poorly solvated in the dipolar aprotic solvents. On the other hand, owing to the presence of atoms with lone pairs of electrons, their interaction with

*Solvent Effects in Chemistry*, Second Edition. Erwin Buncel and Robert A. Stairs.
© 2016 John Wiley & Sons, Inc. Published 2016 by John Wiley & Sons, Inc.

**TABLE 6.1  Physical Properties of Some Dipolar Aprotic Non-HBD Solvents in Order of Increasing Dipole Moment**

| Name | Formula | bp(°C) | Relative permittivity, $\varepsilon_r$ | Dipole moment, $(10^{-30}\,C\,m)$ | $E_T^N$ | $pK_{auto}$ |
|---|---|---|---|---|---|---|
| 2-Propanone (acetone) | $CH_3COCH_3$ | 56 | 20.6 | 9 | 0.355 | 32.5 |
| Formamide | $HCONH_2$ | 210 | 111 | 11.2 | 0.799 | 16.8 |
| Nitromethane | $CH_3NO_2$ | 101 | 35.9 | 11.9 | 0.481 | 24 |
| N,N-Dimethylacetamide | $CH_3CONMe_2$ | 166 | 37.8 | 12.4 | 0.401 | 24 |
| N,N-Dimethylformamide (DMF) | $HCONMe_2$ | 153 | 36.7 | 13 | 0.404 | 29.4 |
| N-Methylpyrrolidin-2-one (NMP) | $CH_3NCOCH_2CH_2CH_2$ | 202 | 32.2 | 13.6 | 0.355 | 24.2 |
| Acetonitrile | $CH_3CN$ | 81.6 | 35.9 | 11.8 | 0.46 | ≥33.3 |
| Dimethyl sulfoxide (DMSO) | $CH_3SOCH_3$ | 189 | 46.5 | 13.5 | 0.444 | 33.3 |
| N,N-Dimethylpropylene urea (DMPU) | $CH_2CH_2CH_2N(CH_3)CONCH_3$ | 230 | 36.1 | 14.1 | 0.352 | – |
| Sulfolane | $(CH_2)_5SO_2$ | 287 | 43.3 | 16 | 0.41 | 25.5 |
| Propylene carbonate (PC) | $OCH(CH_3)CH_2OCO$ | 242 | 64.9 | 16.5 | 0.491 | – |
| Hexamethylphosphoric triamide (HMPT) | $(Me_2N)_3PO$ | 233 | 29.6 | 18.5 | 0.315 | 20.6 |

cations will be quite effective. It is this characteristic inability to solvate anions, contrasting with cations, which is used to advantage in reactivity studies to be described later.

Not all dipolar aprotic solvents are organic. In nucleophilic substitution reactions at aliphatic carbon centers, liquid ammonia behaves like a typical dipolar aprotic solvent, albeit with some notable mechanistic differences. These have been explored in some depth by Michael Page and his group (Ji *et al.*, 2010, 2012). Solvolysis rates of substituted benzyl chloride in $LNH_3$ at 25°C showed little or no dependence on ring substituents with Hammett rho values practically zero, in accordance with no significant charge being developed on the benzylic carbon in the transition state, that is, a concerted $S_N2$ mechanism, Scheme 1.

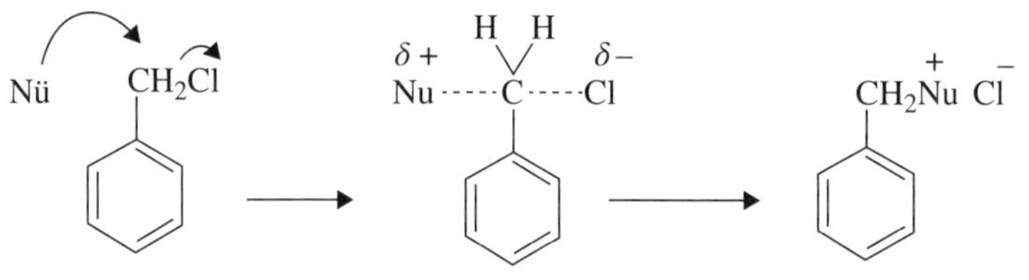

**SCHEME 1**

On the other hand, nucleophilic aromatic substitution in the $LNH_3$ proceeds/gives evidence of an isolatable Meisenheimer complex intermediate, that is, $S_N(AR)$, Scheme 2.

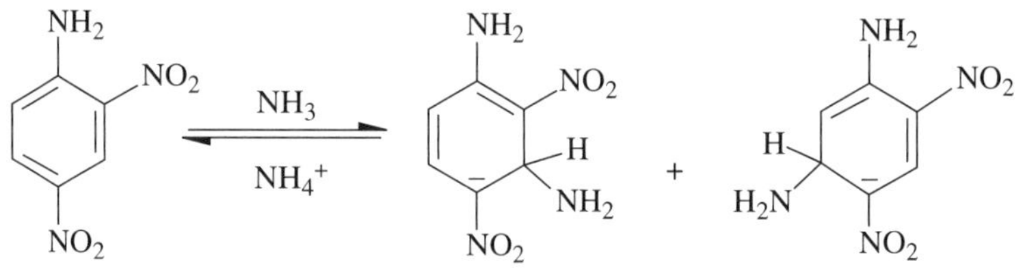

**SCHEME 2**

A dipolar aprotic solvent *par excellence* is dimethyl sulfoxide (DMSO), and most representative studies of solvent effects on reactivities have included DMSO (Buncel and Wilson, 1977). Accordingly, in the following account, DMSO will be highlighted as the prototypical dipolar aprotic solvent, though comparisons with other dipolar aprotic solvents will also be presented.

## 6.2 ACIDITIES IN DMSO AND THE H-SCALE IN DMSO–$H_2$O MIXTURES

DMSO has been found to be highly suitable for study of ionizations of extremely weak acids. Its conjugate base, $CH_3SOCH_2^-$, or *dimsyl anion*, is stable in DMSO, and the glass electrode can be used for potentiometric measurement in DMSO.

**TABLE 6.2  Equilibrium Acidities in Dimethyl Sulfoxide and in Water[a]**

| Acid | p$K_a$(H$_2$O) | p$K_a$(DMSO) |
|---|---|---|
| F$_3$CSO$_3$H | −14 | 0.3 |
| HBr | −9 | 0.9 |
| HCl | −8 | 1.8 |
| CH$_3$SO$_3$H | −0.6 | 1.6 |
| HF | 3.2 | 15 |
| PhCO$_2$H | 4.25 | 11.1 |
| CH$_3$CO$_2$H | 4.75 | 12.3 |
| PhNH$_3$$^+$ | 4.6 | 3.6 |
| PhSH | 6.61 | 10.3 |
| NH$_4$$^+$ | 9.2 | 10.5 |
| CH$_3$NO$_2$ | 10 | 17.2 |
| PhOH | 10.2 | 18 |
| CH$_2$(CN)$_2$ | 11 | 11 |
| F$_3$CCH$_2$OH | 12.4 | 23.6 |
| CH$_3$OH | 15.5 | 29 |
| H$_2$O | 15.7 | 32 |

[a] Bordwell (1988) and Stewart (1985).

Thus, an acidity scale has been established in DMSO and Table 6.2 records some p$K$(DMSO) data together with the corresponding p$K$(H$_2$O) results (Bordwell, 1988). It is noteworthy that for H$_2$O, p$K$(DMSO) is of the order of 32, that is, it is an extremely weak acid! This, of course, corresponds to OH$^-$ being an extremely strong base in DMSO. Interestingly, relative acidities in DMSO resemble those in the gas phase more than in aqueous solution, pointing to the all-important influence of hydrogen bonding in an aqueous medium (Pellerite and Brauman, 1980; Taft and Bordwell, 1988).

DMSO–H$_2$O mixtures also impart a strong basicity to the hydroxide ion. Quantitatively, this is expressed by means of the $H_-$ acidity function, that is,

$$H_- = -\log\left(\frac{a_{H^+}\, f_{A^-}}{f_{HA}}\right) \qquad (6.1)$$

As in the case of $H_0$ (see Chapter 3), in dilute aqueous solution $H_- = \text{pH}$, since in that state $a_{H^+} = c_{H^+}$ and $f_{A^-} = f_{HA} = 1$. A selection of $H_-$ data in DMSO–H$_2$O, and in several other aqueous dipolar aprotic solvent mixtures, is given in Table 6.3 (Buncel, 1975b). The extremely sharp rise of $H_-$ as the last small amounts of water are removed is quite striking. A physical interpretation of this can be given in terms of hydrogen bonding between DMSO and H$_2$O molecules, as illustrated:

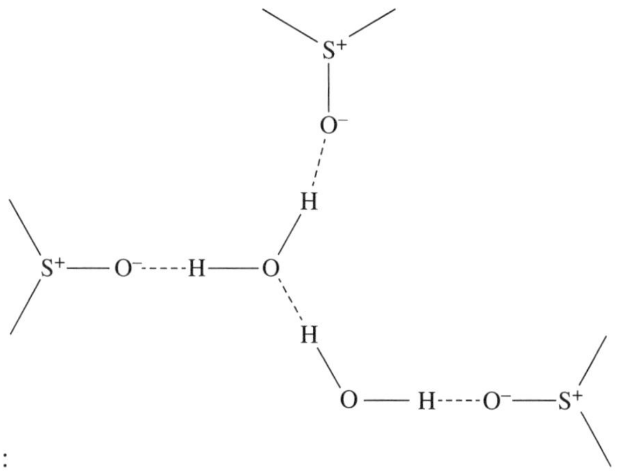

**TABLE 6.3  Selected $H_-$ Data for Aqueous Binary Mixtures with Several Dipolar Aprotic Solvents, Each with 0.011M Tetramethylammonium Hydroxide[a]**

| Mole % dipolar aprotic component | Aqueous pyridine | Aqueous tetramethylene sulfone | Aqueous dimethyl formamide | Aqueous dimethyl sulfoxide |
|---|---|---|---|---|
| 20 | 13.75 | 13.22 | 14.20 | 14.48 |
| 40 | 14.76 | 14.25 | 15.75 | 16.50 |
| 60 | 15.31 | 15.56 | 17.34 | 18.50 |
| 80.78 | | | | 20.68 |
| 90.07 | | | | 21.98 |
| 99.59 | | | | 26.59 |

[a] Buncel (1975b).

This H bonding effectively removes any free $H_2O$ molecules from the bulk solvent, and hence from solvating $HO^-$, which then becomes correspondingly more effective in removing a proton from a solute in an equilibrium or rate process.

## 6.3  USE OF THERMODYNAMIC TRANSFER FUNCTIONS

Approaching solvent effects from the viewpoint of thermodynamic transfer functions allows one to examine in a systematic manner the outcome of medium change, from a protic to a dipolar aprotic reaction medium, in terms of structure and charge distribution in reactants, transition states, and products (Buncel and Wilson, 1979, 1980).

The free energy of activation for a reaction in a particular standard solvent **O**, and in a solvent **S**, may be expressed by Equations 6.2 and 6.3 where T refers to the

$$\Delta G_\mathbf{O}^\ddagger = G_\mathbf{O}^\mathrm{T} - G_\mathbf{O}^\mathrm{R} \tag{6.2}$$

$$\Delta G_\mathbf{S}^\ddagger = G_\mathbf{S}^\mathrm{T} - G_\mathbf{S}^\mathrm{R} \tag{6.3}$$

$$\Delta G_{\text{S}}^{\ddagger} - \Delta G_{\text{O}}^{\ddagger} = \left(G_{\text{S}}^{\text{T}} - G_{\text{O}}^{\text{T}}\right) - \left(G_{\text{S}}^{\text{R}} - G_{\text{O}}^{\text{R}}\right) \tag{6.4}$$

transition state and R to the reactants. Simple subtraction yields Equation 6.4. If the standard free energy of transfer between solvents **O** and **S** is defined according to Equation 6.5, then Equation 6.6 can be derived from Equation 6.4. A pictorial representation of these relationships is shown in Figure 3.7.

$$\delta G_{\text{tr}}^{\text{i}} = G_{\text{S}}^{\text{i}} - G_{\text{O}}^{\text{i}} \tag{6.5}$$

$$\delta G_{\text{tr}}^{\text{T}} = \delta G_{\text{tr}}^{\text{R}} + \Delta G_{\text{s}}^{\ddagger} - \Delta G_{\text{O}}^{\ddagger} = \delta G_{\text{tr}}^{\text{R}} + \delta \Delta G^{\ddagger} \tag{6.6}$$

From Equation 6.6 it is apparent that $\delta G_{\text{tr}}^{\text{T}}$ can be evaluated from calculated values of the transfer free energies of stable solute species, $\delta G_{\text{tr}}^{\text{R}}$, in conjunction with the measured kinetic activation parameters, $\delta \Delta G^{\ddagger}$. The required transfer free energies $\delta G_{\text{tr}}^{\text{R}}$ can readily be obtained from activity coefficient measurements using Equation 6.7 in which $\gamma$ refers to solute activity coefficients in the different solvents ($\gamma_{\text{S}}$ and $\gamma_{\text{O}}$ are referred to the same standard state in solvents

$$\delta G_{\text{tr}}^{\text{i}} = -RT \ln \frac{\gamma_{\text{S}}}{\gamma_{\text{O}}} \tag{6.7}$$

S, O, or any other medium). Methods used to obtain these activity coefficients have included vapor pressure, solubility, and distribution coefficient measurements.

By analogy it is apparent that for the equilibrium situation the transfer free energy relationship will be given by Equation 6.8 where $\delta G_{\text{tr}}^{\text{P}}$ is the transfer energy of the

$$\delta \Delta G_{\text{tr}} = \delta G_{\text{tr}}^{\text{P}} - \delta G_{\text{tr}}^{\text{R}} \tag{6.8}$$

products and $\delta \Delta G_{\text{tr}}$ is the difference in the standard free energies of reaction between the two solvents.

Transfer functions can also be defined for the other thermodynamic state functions. Since enthalpy changes are often conveniently measurable, the transfer enthalpy, $\delta H_{\text{tr}}^{\text{i}}$, is perhaps the most widely used function. From the second law of thermodynamics, the transfer entropy function is given by Equation 6.9.

$$\delta G_{\text{tr}}^{\text{i}} = \delta H_{\text{tr}}^{\text{i}} - T \delta S_{\text{tr}}^{\text{i}} \tag{6.9}$$

Thus, if both transfer free energies and enthalpies are available it should be possible to achieve complete dissection of the effect of solvent on the various thermodynamic parameters.

Other potentially useful transfer functions are the transfer heat capacity, $\delta C_{\text{tr}}^{\text{i}}$, the transfer volume, $\delta V_{\text{tr}}^{\text{i}}$, and the transfer internal energy, $\delta U_{\text{tr}}^{\text{i}}$ (Abraham, 1974). At present, measurements of these transfer functions are available for only a few systems, but they can in certain instances be predicted using scaled-particle theory (Desrosiers and Desnoyers, 1976).

**TABLE 6.4** Free Energies of Transfer of Ions ($\delta \Delta G_{tr}$) From Water To Nonaqueous Solvents At 25°C (molar Scale, in Kcal Mol$^{-1}$)[a]

| Ion | $\delta \Delta G_{tr}$ | | | | | | | |
|---|---|---|---|---|---|---|---|---|
| | MeOH | HCONH$_2$ | NMeF | DMF | DMSO | Me$_2$CO | PC | MeCN |
| H$^+$ | 2.6 | — | — | −3.4 | −4.5 | −0.7 | — | 11.1 |
| Li$^+$ | 1.0 | −2.3 | −3.5 | −5.3 | −3.5 | — | 5.3 | 7.1 |
| Na$^+$ | 2.0 | −1.9 | −1.9 | −2.5 | −3.3 | — | 2.6 | 3.3 |
| K$^+$ | 2.4 | −1.5 | −2.0 | −2.3 | −2.9 | 0.7 | 0.8 | 1.9 |
| Rb$^+$ | 2.4 | −1.3 | −1.8 | −2.4 | −2.6 | 0.5 | −1.3 | 1.6 |
| Cs$^+$ | 2.3 | −1.8 | −1.7 | −2.2 | −3.0 | 0.4 | −3.5 | 1.2 |
| Ag$^+$ | 1.8 | −3.7 | — | −4.1 | −8.0 | 1.5 | 3.3 | −5.2 |
| Tl$^+$ | 1.0 | — | — | −2.8 | −6.0 | — | 2.0 | 2.2 |
| Et$_4$N$^+$ | 0.2 | — | — | −2.0 | −1.2 | — | — | −2.1 |
| Ph$_4$As$^+$ | −5.6 | −5.7 | — | 9.1 | −8.8 | −7.1 | −8.5 | −7.8 |
| Cl$^-$ | 3.0 | 3.3 | 4.9 | 11.0 | 9.2 | 14.0 | 10.1 | 10.1 |
| Br$^-$ | 2.7 | 2.7 | 3.6 | 7.2 | 6.1 | 10.5 | 7.8 | 7.6 |
| I$^-$ | 1.6 | 1.8 | — | 4.5 | 3.2 | 6.7 | 4.6 | 4.5 |
| N$_3^-$ | 2.5 | 2.9 | — | 8.2 | 5.7 | 10.9 | 7.1 | 7.0 |
| ClO$_4^-$ | 1.4 | — | 0.4 | — | — | 3.6 | — | 1.1 |
| OAc$^-$ | 3.7 | — | – | 14.8 | 11.1 | — | — | 13.4 |
| BPh$_4^-$ | −5.6 | −5.7 | – | −9.1 | −8.8 | −7.1 | −8.5 | −7.8 |

[a] Data from Cox (1973): calculated using the assumption, $\delta \Delta G_{tr}(Ph_4As^+) = \delta \Delta G_{tr}(BPh_4^-)$. NMeF, N-methylformamide; PC, propylene carbonate.

On the other hand, use of activity coefficients of transfer, $^0\gamma^S$, instead of free energies of transfer, is an equivalent approach for examination of medium effects on reaction rates and equilibria. Parker has emphasized this alternative approach and the reader is referred to his writings, especially for medium effects in bimolecular nucleophilic substitution (Parker, 1969).

In Table 6.4 are presented some of the currently available $\delta \Delta G_{tr}$ data for transfer of ionic species from water to various dipolar aprotic solvents (Cox, 1973). The most noticeable feature of these data is the striking increase in the free energy of small anions with high charge density on transfer to the dipolar aprotic solvents. This is consistent with the large rate increase observed in the reactions involving these anions. If the anion is small, it should be a strong hydrogen bond acceptor in protic solvents. Such interactions are absent in DMSO, which accounts for these anions being much less solvated in DMSO than in water. That could also explain the observation (Wooley and Hepler, 1972) that the ionization product of water decreases rapidly with increasing DMSO content in mixtures of DMSO and water. In contrast, the free energies for large polarizable anions usually decrease on transfer to dipolar aprotic solvents, as strikingly illustrated for the case of BPh$_4^-$.

The results for the free energies of transfer of cations do not present as uniform a picture in the sense that the solvents acetone, propylene carbonate, and acetonitrile do not display the same behavior as solvents such as DMSO and DMF. It is clear, however, that cations are better solvated in the latter solvents than in water.

## 6.4   CLASSIFICATION OF RATE PROFILE-MEDIUM EFFECT REACTION TYPES

By combination of thermodynamic and kinetic measurements one can obtain values for $\delta G_{tr}^{T}$, which in conjunction with $\delta G_{tr}^{R}$ values can be used to pinpoint the cause of rate change with changing solvent. A number of different types can be envisaged. The $\delta G_{tr}^{T}$, $\delta G_{tr}^{R}$ terms can be positive (destabilization), negative (stabilization), or zero (no effect). When both terms have the same sign we call it a *balancing* situation, and with opposite sign a *reinforcing* situation. The rate effects are expected to be largest in the reinforcing situation and smallest in the balancing situation.

In Table 6.5 we summarize the various possibilities and identify the reaction types in terms of positive (negative), initial-state (transition-state) control. For example, case 3 can be described as a *positive transition-state control* reaction type, and so on. Not all of these situations have been observed so far. It would be of interest to design appropriate reactions to complete this classification.

**TABLE 6.5   Transfer Free Energies of Reactants ( $\delta G_{tr}^{R}$ ) and Transition States ( $\delta G_{tr}^{T}$ ) and Solvent Effects On Reaction Rates. Classification of Reaction Types**[a]

| Case | $\delta G_{tr}^{R}$ | $\delta G_{tr}^{T}$ | Effect on rate[b] | Reaction type |
|------|------|------|------|------|
| 1 | − | − | +, 0, or − | Balanced |
| 2 | + | − | + | Positively reinforced |
| 3 | 0 | − | + | Positive transition-state control |
| 4 | − | 0 | − | Negative initial-state control |
| 5 | + | 0 | + | Positive initial-state control |
| 6 | 0 | 0 | 0 | Solvent independent |
| 7 | − | + | − | Negatively reinforced |
| 8 | + | + | +, 0, or − | Balanced |
| 9 | 0 | + | − | Negative transition-state control |

[a] Buncel and Wilson (1980).
[b] The plus sign refers to rate acceleration, the minus to rate retardation, and zero to no effect.

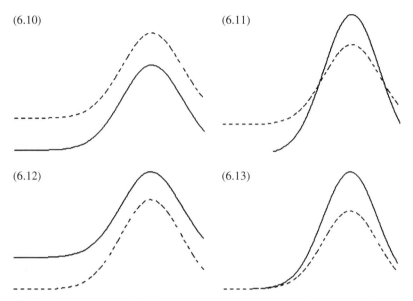

**FIGURE 6.1** Reaction profile illustrations of some representative reactions for changes from protic (solid curves) to dipolar-aprotic (dashed) media, showing enthalpy as the ordinate. Using the classification in Table 6.5, cases 6.10 and 6.12 represent balancing situations, and 6.11 represents positively reinforced and 6.13 represents negative transition-state control. Source: Buncel and Wilson (1979) Redrawn by permission of the American Chemical Society.

$$HO^- + D_2 \quad (H_2O \text{ versus aq. DMSO}) \tag{6.10}$$

$$\text{(p-nitrophenoxide)} + CH_3I \quad (\text{MeOH versus DMF}) \tag{6.11}$$

$$SCN^- + \text{(1-iodo-2-nitro-4-methylbenzene)} \quad (\text{MeOH versus DMF}) \tag{6.12}$$

$$\text{(MeOH versus DMF)} \qquad (6.13)$$

In Figure 6.1 are shown a few examples of reactions (Buncel and Symons, 1976; Haberfield, 1971) of these types, some of which will be referred to again subsequently. It may be noted that only in example (6.13) is the ground-state effect zero. In the general case, however, the rationalization of medium effects on reaction rates requires knowledge of the pertinent transfer function data for the reactants; these are now becoming increasingly available for neutral molecules as well as ionic reagents. In the case of transfer functions for ionic reagents it has been customary to estimate thermodynamic quantities for single ions by use of *extrathermodynamic assumptions*, such as $\delta\Delta G_{tr}(Ph_4P^+) = \delta\Delta G_{tr}(BPh_4^-)$. These methods are open to criticism, but in certain instances they can be avoided by use of appropriate thermochemical cycles (Gold, 1976).

Analogous considerations are applicable in principle to certain excited-state processes. The blue shift of $n \rightarrow \pi^*$ transitions on going to more polar solvents has been analyzed in this manner by Haberfield *et al.* (1977) for a number of ketones and azo compounds. According to the terminology given in Table 6.5, with substitution of the excited state for the transition state, the commonly observed blue shift for ketones on transfer from a nonpolar solvent to a hydrogen-bonding solvent falls under the category of a *negatively reinforced* type of process.

## 6.5 BIMOLECULAR NUCLEOPHILIC SUBSTITUTION

The relative inability of dipolar aprotic solvents to interact effectively with (solvate) anions, especially small (hard) anions in which the negative charge is located on an electronegative atom, results in very large rate accelerations in these non-HBD solvents relative to protic ones. According to the terminology in Table 6.5, these processes can be classified as *positively reinforced* since along with destabilization of the reactants there is stabilization of the transition state due to charge dispersal (Hughes–Ingold rules; Ingold, 1969).

These ideas are nicely illustrated by the results in Table 6.6, corresponding to the four reactions given later (Reichardt *et al.*, 1988, p. 218). The first reaction is a simple $S_N2$ type of halide interchange process, the second is similar but involves azide ion as a very effective nucleophile. The last two are aromatic nucleophilic substitution ($S_NAr$) reactions and are shown as proceeding by way of *sigma complex intermediates* (Buncel *et al.*, 1995). Note that Reaction (6.17), which

**TABLE 6.6    Relative Rates of the $S_N2$ Anion–molecule Reactions A and B and of the $S_N$Ar Reactions C and D in Protic and Polar Non-HBD Solvents At 25°C$^a$**

| | $\log(k_2^{\,solvent}/k_2^{\,MeOH})$ for reaction | | | |
|---|---|---|---|---|
| Solvents | 6.14 | 6.15 | 6.16 | 6.17 |
| **Protic solvents** | | | | |
| $CH_3OH$ | 0 | 0 | 0 | 0 |
| $H_2O$ | 0.05 | 0.8 | — | — |
| $CH_3CONHCH_3$ | — | 0.9 | — | — |
| $HCONH_2$ | 1.2 | 1.1 | 0.8 | — |
| $HCONHCH_3$ | 1.7 | — | 1.1 | — |
| **Dipolar non-HBD solvents** | | | | |
| $(CH_2)_4SO_2$ (sulfolane) | — | 2.6 | 4.5 | — |
| $CH_3NO_2$ | 4.2 | — | 3.5 | 0.8 |
| $CH_3CN$ | 4.6 | 3.7 | 3.9 | 0.9 |
| $CH_3SOCH_3$ | — | 3.1 | 3.9 | 2.3 |
| $HCON(CH_3)_2$ | 5.9 | 3.4 | 4.5 | 1.8 |
| $CH_3COCH_3$ | 6.2 | 3.6 | 4.9 | 0.4 |
| $CH_3CON(CH_3)_2$ | 6.4 | 3.9 | 5.0 | 1.7 |
| $N$-Methylpyrrolidinone | 6.9 | — | 5.3 | — |
| $[(CH_3)_2N]_3PO$ | — | 5.3 | 7.3 | — |

$^a$ Reichardt (2003, p. 249).

involves a neutral nucleophile, is much less sensitive to solvent change than the former three, which all involve anionic nucleophiles; this is in accordance with expectations.

$$Cl^- + CH_3 - I \xrightarrow{k_2} Cl - CH_3 + I^- \tag{6.14}$$

$$N_3^- + CH_3CH_2CH_2CH_2 - Br \xrightarrow{k_2} CH_3CH_2CH_2CH_2 - N_3 + Br^- \tag{6.15}$$

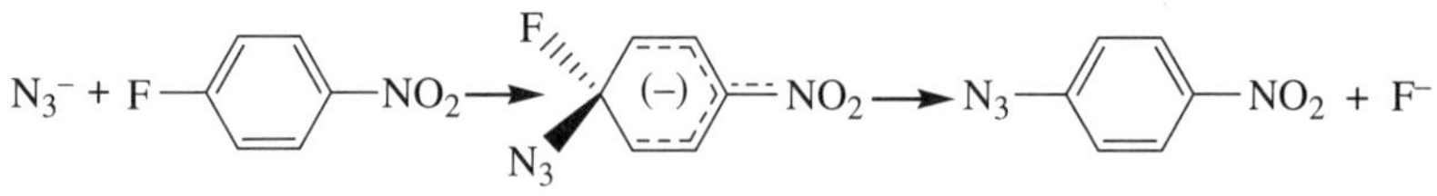

(6.16)

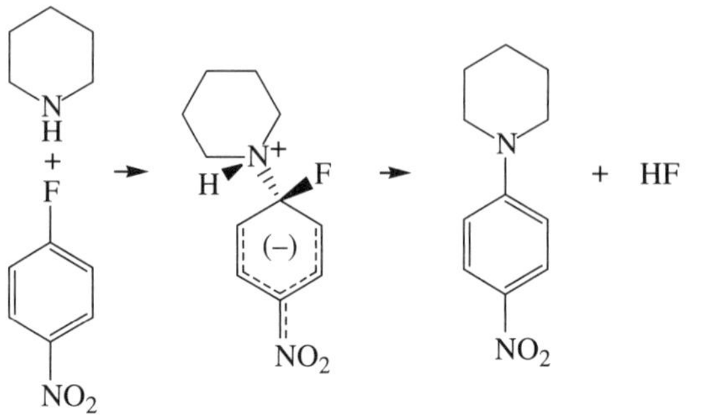

(6.17)

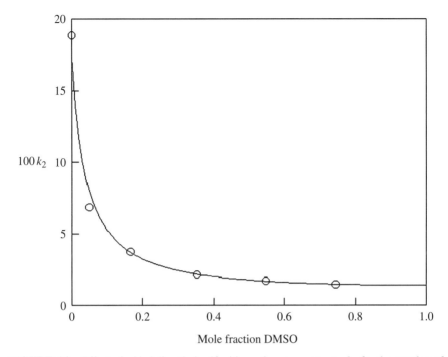

**FIGURE 6.2**  Effect of added dimethyl sulfoxide on the rate constants, $k_2$, for the reaction of benzyl chloride with the 9-cyanofluorenyl anion in ethanolic media at 35.7°C. Data from Bowden and Cook (1968).

The most important practical consequence of the huge rate accelerations effected by dipolar aprotic (non-HBD) solvents is that many SN2 SNA2 type reactions have become feasible under mild laboratory conditions, which was not the case as long as only common hydroxylic solvents were available. Another important element is the *reversal in nucleophilic reactivity* of small anions, which can be effected by this solvent change and which also results from the contrasting solvation capabilities of the two kinds of solvent.

Not all nucleophilic substitution reactions exhibit an increase in rate on going to a dipolar aprotic solvent from a protic one. The reaction of 9-cyanofluorenyl anion (9-CFA) with benzyl chloride (6.18) is a case in point (Bowden and Cook, 1968). The rate data plotted in Figure 6.2 show that adding DMSO to an ethanolic medium causes a moderate retardation in the rate for this reaction. This result is understandable, however, on the basis that the large charge-delocalized 9-CFA anion would be strongly stabilized by the polarizable DMSO (i.e., $\delta\Delta G_{tr}$ negative), and that stabilization of the transition state (increased charge distribution) would not be sufficient to overcome this. According to Table 6.5 this system is denoted as subject to *negative initial-state control*.

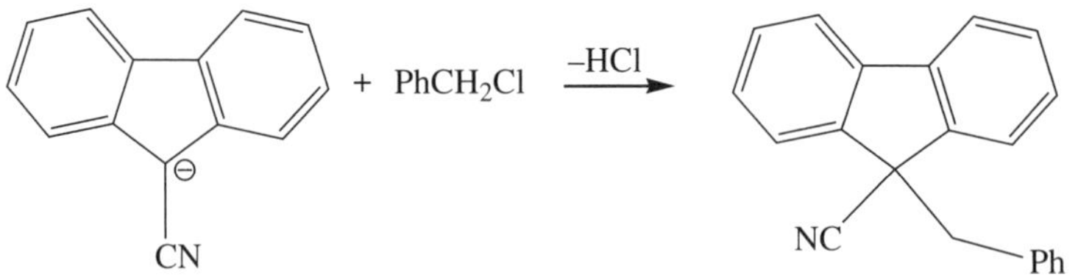

$$\text{[fluorene structure with } CN \text{] } + \text{ PhCH}_2\text{Cl} \xrightarrow{-\text{HCl}} \text{[fluorene product with NC and Ph]} \qquad (6.18)$$

## 6.6   PROTON TRANSFER

Some of the largest rate enhancements due to solvent change have been observed for proton transfer processes. For example, the rate of the methoxide-catalyzed racemization of 2-methyl-3-phenylpropionitrile, $PhCH_2CH(CH_3)CN$, is increased by a factor of $10^8$ on going from methanol to DMSO (Cram et al., 1960). Racemization processes generally involve rate-determining proton abstraction followed by fast reaction with a solvent to yield a planar carbanion. Fast protonation of this carbanion results in the formation of the two enantiomers in equal proportion.

Isotopic hydrogen exchange follows a similar mechanism, as shown in Equations 6.19 and 6.20 for the case of a protium-containing substrate undergoing exchange with

$$R-H+B^- \xrightarrow{\text{slow}} R^- + HB \qquad (6.19)$$

$$R^- + D-OD \xrightarrow{\text{fast}} R-D+DO^- \qquad (6.20)$$

deuterium-containing solvent. Isotopic hydrogen exchange has been investigated in a great variety of C–H–containing substrates, including ketones, nitro compounds, nitriles, halides, sulfoxides, and aromatic hydrocarbons (Buncel and Dust, 2003; Buncel et al., 1990). Generally, the isotopic exchange is facilitated when the negative charge in the carbanion can be delocalized through resonance, as shown for the carbanions derived from acetone, nitromethane, and toluene:

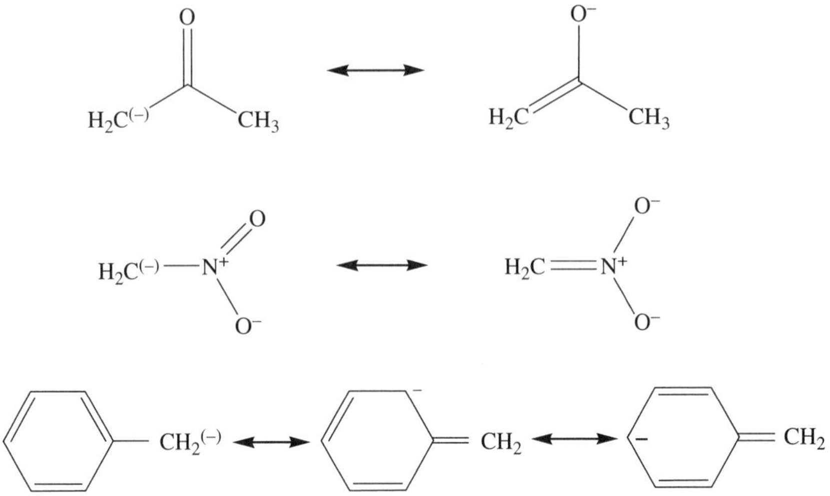

In the case of the trihalomethanes, which also undergo deuteroxide-ion-catalyzed H/D exchange (Symons and Clermont, 1981), carbanion stabilization occurs through inductive electron withdrawal by the electronegative halogens. However, that this stabilizing effect is much smaller quantitatively than the conjugative effect is seen by the fact that chloroform, for example, is a weaker acid than $CH(NO_2)_3$ by ca. $20\,pK$ units. Interestingly, for $CF_3D/HO^-(0.01M)$ at 49°C, the rate of exchange increases by 6 orders of magnitude on going from water to 70 mol% DMSO (Symons *et al.*, 1981).

An interesting case of isotopic hydrogen exchange is that in sulfoxides. Thus, deuterated dimethyl sulfoxide (widely used in NMR studies) can be prepared by allowing $CH_3SOCH_3$ (95% v/v) to equilibrate with $DO^-/D_2O$ (5% v/v) at 100°C several times and removing the exchanged HDO through distillation (Buncel *et al.*, 1965). Apparently, the basicity of hydroxide (deuteroxide) ion in DMSO-rich media is enhanced to a sufficient degree to allow facile proton abstraction to occur:

$$CH_3\!-\!\overset{\overset{\displaystyle O}{\|}}{S}\!-\!CH_3 \xrightarrow{\ DO^-\ } HOD + CH_3\!-\!\overset{\overset{\displaystyle O}{\|}}{S}\!-\!CH_2^- \xrightarrow{\ DOD\ } DO^- + CH_3\!-\!\overset{\overset{\displaystyle O}{\|}}{S}\!-\!CH_2D \qquad (6.21)$$

## 6.7   D$_2$–HO$^-$ EXCHANGE

The isotopic exchange between molecular hydrogen (dihydrogen) and a hydroxylic solvent under base catalysis poses a rather intriguing mechanistic problem. The two main mechanisms that had been proposed (Symons and Buncel, 1972) involve rate-determining proton transfer (Eqs. 6.22 and 6.23) or formation of an addition complex (Eqs. 6.24 and 6.25):

$$HO^- + D\!-\!D \xrightarrow{\ slow\ } HOD + D^- \qquad (6.22)$$

$$D^- + HOD \xrightarrow{\ fast\ } DH + HO^- \qquad (6.23)$$

$$HOH\text{---}O^- + D\!-\!D \xrightarrow{\ slow\ } HOH + (HO\!-\!D\!-\!D)^- \qquad (6.24)$$
$$\underset{\displaystyle H}{|}$$

$$[HO\!-\!D\!-\!D]^- + HOH \xrightarrow{\ fast\ } HOD + DH + HO^- \qquad (6.25)$$

The rate constant for exchange increases by ca. $10^4$ on changing the medium composition from purely aqueous to 99.5 mole % DMSO at 65°. It is significant that this increase in rate is considerably less than observed in many other reactions for

**TABLE 6.7  Enthalpies of Transfer (kcal Mol$^{-1}$) for Reactants and for the Transition State of the D$_2$–HO$^-$ Exchange Process in the DMSO–H$_2$O System$^a$**

| Mol % DMSO | $\delta\Delta H_{tr}^{D2}$ | $\delta\Delta H_{tr}^{HO^-}$ | $\delta\Delta H_{tr}^{R}$ | $\Delta H^\dagger$ | $\delta\Delta H^\dagger$ | $\delta\Delta H_{tr}^{T}$ |
|---|---|---|---|---|---|---|
| 0 | 0 | 0 | 0 | 24.0 | 0 | 0 |
| 10 | 1.2 | −0.8 | 0.4 | — | — | — |
| 20.2 | 2.1 | +3.7 | 5.8 | 21.2 | −2.8 | 3.0 |
| 40.1 | 2.6 | 10.1 | 12.7 | 17.7 | −6.3 | 6.4 |
| 59.0 | 2.2 | 13.8 | 16.0 | 16.4 | −7.6 | 8.4 |
| 77.9 | 2.1 | 16.0 | 18.1 | 16.9 | −7.1 | 11.0 |
| 87.5 | 2.1 | 16.9 | 19.0 | 16.8 | −7.2 | 11.8 |
| 96.9 | 2.1 | 17.5 | 19.6 | 18.1 | −5.9 | 13.7 |

$^a$ Buncel and Wilson (1980).

which the medium effect has been evaluated. Analysis in terms of thermodynamic and kinetic transfer functions gives information about the origin of the observed medium effect.

In Table 6.7 are presented the relevant data, derived from measured values of enthalpies of activation and available literature data for the enthalpies of transfer of hydrogen and hydroxide ion (Buncel and Symons, 1976). It is apparent that there is a close parallel between the transfer enthalpies for the reactants, $\delta\Delta H_{tr}^{R}$, and those for the transition state, $\delta\Delta H_{tr}^{T}$. There is some similarity with alkaline ester hydrolysis, in which case the transition-state enthalpy transfer is also endothermic, though not nearly to the same degree as in the present system. Evidently, the destabilization of HO$^-$ (noting that $\delta\Delta H_{tr}(HO^-)$ is the major component of $\delta\Delta H_{tr}^{R}$ in DMSO-rich media) is largely retained in the transition state. These results point to a rate-determining transition state with considerable charge localization on an electronegative atom. Any proposed mechanism must be in accordance with this conclusion. The relative merits of various mechanisms in the light of this finding are considered in detail elsewhere (Buncel and Symons, 1976).

## PROBLEMS

**6.1**  A substance A is miscible in all proportions with a solvent S, forming nearly ideal solutions. The mutual solubilities of A and water at 25°C are 0.0023 mole fraction of A in water, 0.0010 m.f. of water in A. Estimate the Gibbs energy of transfer of A from S to water.

**6.2**  The same substance A undergoes a reaction $A \rightleftharpoons A^\ddagger \rightarrow$ products, which proceeds 30 times as fast at 25° in water as in S. Estimate the transfer free energy of the activated complex from S to water.

# 7

# EXAMPLES OF OTHER SOLVENT CLASSES

## 7.1 INTRODUCTION

In this chapter we consider a few examples other than the dipolar-aprotic ones treated in Chapter 6. We have chosen solvents somewhat arbitrarily, perhaps because we have ourselves had occasion to use them, or because they exhibit particular properties that make them useful for special purposes. Sometimes mere inertness is all that is required, as in the case of solvents for chromyl chloride, mentioned in Chapter 2. (As it is a very active oxidant, it inflames immediately with diethyl ether, for instance.) The properties required may include the ability to dissolve a wide variety of substances, to sustain strong acidity, basicity, or oxidizing or reducing strength, a high or low boiling point, long liquid range, high or low relative permittivity, or low or (rarely) high viscosity. We continue by considering examples of solvents under three headings: acidic, basic, and chiral.

## 7.2 ACIDIC SOLVENTS

Acidic solvents may be used for their catalytic effect. Though many reactions are catalyzed by protons available in ordinary solvents, including water, many other reactions require high levels of acidity to proceed at useful rates. These acidities are only attainable if the solvent is either so low in basicity that strong acids can be

*Solvent Effects in Chemistry*, Second Edition. Erwin Buncel and Robert A. Stairs.
© 2016 John Wiley & Sons, Inc. Published 2016 by John Wiley & Sons, Inc.

TABLE 7.1    Substances Soluble in HF with Acid/base Properties

| Acids | Bases (not ordered) |
|---|---|
| Appreciable (enhance acidity of HF): | MF: M=Li, Na, etc.; $NH_4$, Ag, Tl, $CH_3NH_3$, |
| $SbF_5 > AsF_5 > NbF_5 > PF_5 > BF_3 > SnF_4$ | $(CH_3)_4N$, $C_6H_5NH_3$, $(C_6H_5)_2NH_2$, etc. |
| Weak (react with $F^-$): $TiF_4 > VF_5 > MoF_5 >$ | $MF_2$: M=Ca, Sr, (Ba) |
| $WF_6 > ReF_6 > GeF_4 > TeF_6 > SeF_4 > IF_4 >$ | |
| $SiF_4$, $PrF_4$ | |

present in solution, or if it is itself moderately acidic. Acidity can sometimes be increased by the addition of a Lewis acid, as described in Chapter 3.

## 7.2.1    An Acidic Solvent; Hydrogen Fluoride

In spite of the well-known hazards associated with its use,[1] and its chemical activity toward glass and many other materials, hydrogen fluoride has been used as a solvent since it was first prepared in the anhydrous state by Frémy (1856). Its boiling point, 19.5°C, is not inconveniently low, but it is easily removed from a reaction mixture by distillation during workup. HF may be handled in copper, nickel, or Monel vessels, with minimal reaction, though only Pt or Pt/Au alloy does not cause any contamination. Nowadays vessels lined with Teflon® or "pctfe" (poly(chlorotrifluoroethene)) are used.

Pure HF is very strongly acidic ($H_0 \approx -15$). Its autoprotolysis constant is comparable to that of water ($pK_{ip} = 13.7$). No common Brønsted acid is strong in HF (sulfuric, fluorosulfonic, and perchloric acids are nonelectrolytes), but through reactions with Lewis acids, such as

$$PF_5 + 2HF = H_2F^+ + PF_6^-  \qquad (7.1)$$

one can reach large negative values of $H_0$. Table 7.1 lists some substances exhibiting various degrees of acidity and basicity in HF (Dove and Clifford, 1971, p. 174 and *passim*).

Spectroscopic measurements in HF are feasible using windows of quenched (glassy) pctfe (visible only), alumina (synthetic sapphire, UV-visible), or fused silver chloride (IR). For NMR, samples can be sealed in Teflon or pctfe tubing and encased in glass to permit spinning.

The solubilities of inorganic substances in HF resemble those in water, with some marked differences. Most elements are not dissolved, except the active metals, which react with evolution of hydrogen gas. Alkali metal and some alkaline earth salts are soluble, but many react liberating weaker acids in solution. Alkali metal halides evolve the nearly insoluble hydrogen halides. Salts of transition metals are at most

---

[1]**Caution:** HF gas is very irritating to mucous membranes. The liquid quickly penetrates the skin, and causes painful burns and necrosis. Rubber gloves, an efficient hood, and immediately available first-aid supplies are essential.

slightly soluble. Organic solubilities tend to be greater than those of the same substances in water. The most soluble substances are those carrying atoms or groups with EPD ability, such as O, N, S, or C=C.

### 7.2.2 Reactions in Hydrogen Fluoride

Chemistry in HF has been reviewed by Kilpatrick and Jones (1967) and by Dove and Clifford (1971), and particularly in its industrial aspects, by Smith (1994). The properties that make HF useful are its great acidity, its ability to dissolve a variety of organic and inorganic substances, and its resistance to oxidation. It is used in petroleum refining to carry out alkylations, such as the reaction of isobutane with light olefins to produce high-octane, branched C7 and C8 aliphatics that, being insoluble in HF, are readily separated from the solvent/catalyst. In the manufacture of detergents, similar alkylation reactions using HF as solvent and catalyst lead to a high proportion of the linear isomers, desirable because they are biodegradable (but see later, Section 8.3.1).

### 7.2.3 Electrochemistry in Hydrogen Fluoride

Studies of electrical conductivity in HF cited by Kilpatrick and Jones show mobilities of ions somewhat larger than in water. $H^+$ and $F^-$ are more mobile than most ions (cf. $H^+$ and $HO^-$ in water), suggesting a proton-jump (Grotthuss) mechanism. The concentration dependence of the conductance of all but a few electrolytes indicates considerable ion pairing, in spite of the high relative permittivity (83.6 at 0°C). $Hg(CN)_2$, on the other hand, a nonelectrolyte in most solvents, gives conducting solutions. Electrolysis of solutions of alkali fluorides in HF yields fluorine at the anode. HF has great affinity for water. If the solvent is not dry, the evolved fluorine is contaminated with $OF_2$, $O_2$, and ozone, but continued electrolysis is effective in removing water. Hydrocarbons and certain derivatives in HF solution are perfluorinated, at a less anodic potential than is required to liberate fluorine. For example $n$-hexane is converted to perfluoro-$n$-hexane according to Equation 7.2;

$$C_6H_{14} + 28F^+ \rightarrow C_6F_{14} + 14HF + 28e^-. \tag{7.2}$$

Except by electrolysis, HF is not easily oxidized. It should be a suitable solvent for strong oxidants. Dove and Clifford (1971) describe some such uses, but report experimental difficulties, involving reaction with reducing impurities, and attack by such very reactive solutes as rhenium heptafluoride and xenon hexafluoride on pctfe containers. Tables of electrochemical potentials are also presented.

### 7.2.4 Some Other Acid Solvents

Recalling the idea of Usanovich (Section 3.14) that oxidation and acidity have similar aspects, it is not surprising that oxidation of hydrocarbons can be carried out in superacid media. Main group elements in higher oxidation states, such as

Hg(II) in these media, can oxidize the lower hydrocarbons, but with poor selectivity owing to the formation of free radicals. Hashiguchi *et al.* (2014) describe the successful oxidation of methane, ethane, and propane in trifluoro-acetic acid to trifluoroacetate esters, using thallium (III) and lead (IV) trifluoro-acetates. Yields of the esters were better than 75% in most cases. The products from ethane oxidation typically consisted of the ethyl ester and the diester of ethanediol in about 3:1 ratio. From propane the main product was the ester of 2-propanol.

## 7.3   BASIC SOLVENTS

As was noted in Chapter 3, the presence of one or more unshared pairs of electrons confers both Brønsted and Lewis basicity, and nucleophilicity, on the molecule. This may be manifested as the ability to solvate cations, to accept hydrogen bonds, to sta-bilize normal species or activated complexes by association with positive charges or electron-poor regions. Values of the empirical parameters that purport to measure basicity, $\beta$, $B_j$, $DN$, $pK_{BH^+}$ (aq), and so on, for different basic solvents accordingly differ even in their order of strengths. The hard/soft classification does not explain all the differences.

### 7.3.1   A Basic Solvent: Ammonia

Organic solvents that are notably basic are almost all nitrogen-containing com-pounds; the simplest of these is ammonia. B. G. Cox (2013, p. 93) describes it as "a typical, but strongly basic, polar aprotic solvent." The physical properties of liquid ammonia are not very different from those of hydrogen fluoride. Its boiling point is lower (-33°C), but not too inconveniently so.[2] Work in the laboratory of M. J. Page (Ji *et al.*, 2010, 2012) demonstrates that it can be handled in glass at temperatures near and above room temperature, with precautions. It can be drawn from the cylinder as liquid or as gas. It is intensely hygroscopic, but is effectively dehydrated by distillation from the blue solution that it forms with sodium. Inorganic substances tend to be less soluble in ammonia than in water, especially salts of 1:2 or higher valence types. Reversals of order are observed where cations of salts form ammines, notably silver and lithium salts. Solubilities of organic substances in ammonia are often greater than in water. The most striking examples of substances with unusual solubilities are the alkali and alkaline earth metals. The alkali metals dissolve without immediate reaction to form intensely blue (dilute) solutions. Water can also dissolve metals, as shown by the transient blue color observed around the cathode when a concentrated NaOH solution is subjected to electrolysis at high

---

[2]**Caution**: Because it has a low boiling point and a relatively large heat of vaporization, and because it wets the skin, ammonia can quickly cause frostbite. It is poisonous by inhalation and requires adequate precautions.

current density. It is the basicity, or rather the low acidity, of ammonia that slows the decomposition. In the presence of ammonium salts, which are acids in ammonia, the reaction is rapid:

$$2NH_4^+ + 2Na \rightarrow 2NH_3 + H_2 + 2Na^+ \qquad (7.3)$$

More concentrated solutions of alkali metals in ammonia (>0.04 mol fraction of metal) have the appearance of liquid gold. The solubility of lithium in ammonia is large (>0.22 mol fraction), and most of the ammonia is bound in complexes: $Li(NH_3)_{4-x}$. The saturated solution consequently has a normal boiling point above room temperature. On standing the alkali metals react to form the amides with evolution of hydrogen, according to Equation 7.4:

$$2M + 2NH_3 \rightarrow 2M^+NH_2^- + H_2 \qquad (7.4)$$

Li and Na react slowly, and the heavier members of the family more quickly, though none so vigorously that they cannot be effectively used. The alkaline earth metals are less soluble. These solutions appear to contain electrons, in the most dilute solutions solvated by ammonia. Stairs (1957) discusses a simple model for the solvated electron, trapped in a potential well created by polarization of the liquid. As the concentration is increased, the electrons become loosely bound to cations, and dimeric species begin to appear. To explain the concentration dependence of the activity, magnetic properties, and conductance of these metal ammonia solutions, Dye (1964) found it necessary to include a variety of species; the solvated metal ion $M^+$ and electron $e^-$, but also the "atom" or ion pair $M^+e^-$, the triplets $e^-M^+e^-$ and $M^+e^-M^+$, and the quadruplet or "dimer" $(M^+e^-)_2$. Feng and Kevan (1980) reviewed theoretical models. At the highest concentrations, the solution has a metallic appearance. Electrons can move among the cation sites as in a metal, and the electrical conductivity of the most concentrated solutions is comparable to that of mercury. The most useful chemical applications of these solutions are as reducing agents, as in the Birch reduction of aromatic hydrocarbons to alicyclics (see, e.g., Pine (1987, pp. 682, 936); Streitwieser et al. (1992, pp. 634–636)).

The acid–base range attainable in ammonia is large, corresponding to a pH range of 27 units or more, depending on temperature. $H_-$ in the pure liquid at $-33°C$ is $\approx 22$; in 0.1 M $KNH_2$ solution it is $\approx 35$. Exchange reactions such as reaction 7.5

$$NH_3 + D_2 \rightleftharpoons NH_2D + HD \qquad (7.5)$$

are catalyzed (e.g., by $KNH_3$), as are similar reactions of aliphatic amines. They are used for industrial production of deuterium (Buncel and Symons, 1986).

Sodium amide is nearly insoluble in ammonia, but potassium amide, which is soluble, reacts with sodium and other insoluble metal amides to form soluble amido complexes, such as $Na(NH_2)_3^{2-}$ and $Al(NH_2)_4^-$. Amphoterism is thus much commoner in the ammonia system of compounds than in the water system (Audrieth and Kleinberg, 1953, p. 81).

Ammonia is a leveling solvent for acids. All Brønsted acids except the weakest are converted to ammonium salts, which are acids in ammonia ("ammono-acids"). Owing to ammonia's rather low relative permittivity, ion pairing occurs, reducing the apparent acidity; the dissociation constants of ammono-acids as diverse as $(NH_4)_2S$ and $NH_4ClO_4$ range from only $9.8 \times 10^{-4}$ to $5.4 \times 10^{-3}$. Nevertheless, solutions of ammonium salts in liquid ammonia will corrode some metals and dissolve many metal oxides. Divers's solution (a saturated solution of ammonium nitrate in ammonia), which is stable to near room temperature, acts on metals much as does aqueous nitric acid.

The solubility, properties in solution, and reactions of a large number of organic compounds in liquid ammonia were reviewed by Smith (1963), and of inorganic substances by Jander (1966) and by Lagowski and Moczygemba (1967) and again by Lagowski (1971) in the Symposium on Non-Aqueous Electrochemistry (Paris, July 1970). The last devotes particular attention to electrochemical properties of both inorganic and organic substances in ammonia.

Ji *et al.* (2010, 2012) discuss a number of types of reactions of organic substances in liquid ammonia, and provide many references to recent and earlier work. They compare the equilibrium constants of phenols and carbonyl-activated carbon acids in ammonia and water, and describe work that is part of an ongoing study of the kinetics of a variety of reactions, including aromatic substitutions and solvolyses in ammonia. They point out that owing to its weakness as an acid and as a hydrogen-bond donor, in many respects ammonia behaves as a dipolar aprotic solvent. It solvates cations strongly, but anions hardly at all (Marcus, 1983, 1985). The low value of the autoprotolysis constant ($\sim 10^{-27}$ at $-33°C$) is chiefly due to the weakness of ammonia as an acid. The mobility of the $NH_4^+$ ion in liquid ammonia is not anomalous (Lagowski, 1971), for the same reason; unlike water, in which the apparent mobilities of $H_3O^+$ and $HO^-$ ions are much enhanced by the facile transfer of protons from $H_3O^+$ to $H_2O$ and from $H_2O$ to $HO^-$ (the Grotthuss mechanism; see Chapter 1).

### 7.3.2  A Basic Solvent: Pyridine

Pyridine is familiar to organic chemists as a reagent, as a reaction medium, and as a component of chromatographic elution solvents. It is relatively easy to purify, though some related substances are not easily removed by fractional distillation. If such impurities are expected to interfere, fractional freezing is recommended. Its physical properties are not inconvenient: a rather long liquid range at ordinary pressure ($-40.7$ to $115.5°C$), a moderate relative permittivity (12.5 at $20°C$), and a viscosity slightly less than that of water.[3]

---

[3]**Caution**: Pyridine is reported to cause gastrointestinal upset and central nervous system depression at high levels of exposure, and to depress sexual activity in men. It can be absorbed through the skin; rubber gloves are recommended. The vapor pressure of pyridine at $20°C$ is 14.5 Torr. The resulting concentration in air would be four times the maximum allowed concentration, $15\,mg\,m^{-3}$, but it is detectable by its strong odour at much lower concentrations.

Inorganic salts are more soluble in pyridine than its relative permittivity would lead one to expect, presumably owing to solvation of the cation through the nitrogen. (The unshared pair of electrons occupies a nonbonding $\sigma$-type orbital, not part of the $\pi$ system of the ring, so is available.) Solubility is favored by a high charge-to-size ratio for the cation, but the reverse for the anion. Pyridine is classified as a borderline base (see earlier, Chapter 3). In a conductivity study (Hantzsch and Caldwell, 1908) it appeared to differentiate the (hard) strong acids; $HI > HNO_3 > HBr > HCl$, but it is not clear that ion pairing was absent. The apparent dissociation constant of perchloric acid in pyridine is $7.55 \times 10^{-4}$. A scheme such as the following (at least) should be assumed:

$$Py + HClO_4 \rightarrow \left[ PyH^+ClO_4^- \right] \rightleftharpoons PyH^+ + ClO_4^- \tag{7.6}$$

Pyridine can sustain a large range of oxidation and reduction potential, depending on the electrolyte present and the electrode material, of more than 3 V. Metals as active as Li, K, and Ba have been successfully electrodeposited from strictly anhydrous pyridine solutions of suitable salts onto Pt or Fe electrodes, or into mercury, though Nigretto and Jozefowicz (1978) in a review of the uses of pyridine as a solvent in analytical, and especially electroanalytical, chemistry, list a number of reagents that reduce pyridine. Pyridine is oxidized by persulfate, but $CrO_3$ dissolves to form Sarett's reagent (House, 1972; Poos et al., 1953), a strong but selective oxidant used to oxidize alcohols to carbonyl compounds (but now superseded by pyridinium chlorochromate). Scriven et al. (1994) describe pyridine as the solvent of choice for acylations (but see later, Section 8.3), and as excellent for dehydrochlorinations, owing to its ability to act as a scavenger for acid.

A well-known use of pyridine as a cosolvent is in the Karl Fischer titration of water in organic solvents, described in Bassett et al. (1978). A solution of iodine and sulfur dioxide in pyridine/methanol or pyridine/cellosolve is fairly stable in the absence of water, but when it is added to a sample of a solvent containing water, reaction 7.7 occurs quantitatively.

$$3C_5H_5N + I_2 + SO_2 + ROH + H_2O \rightarrow 2C_5H_5NH^+I^- + C_5H_5H^+ROSO_3^- \tag{7.7}$$

(R is $CH_3$ or $CH_3OCH_2CH_2$.) The end point is detected by the persistence of the brown color of iodine, or electrometrically.

## 7.4   CHIRAL SOLVENTS

The use of chiral, nonracemic solvents in methods of separation of enantiomers has been reviewed by Eliel et al. (1994). They conclude that separation based on solubility difference between enantiomers in a chiral solvent is unlikely to lead to a practical method of separation. On the other hand, partition between immiscible solvents, one of which is chiral, can lead to separation. In most cases, the degree of separation in one equilibration is very small, but it has been shown to permit at least partial resolution of racemic mixtures by multiple extractions (Bowman *et al.*, 1968), or by the Craig countercurrent method (Leo *et al.*, 1971). For separation in a reasonable number of steps, there seems to be a requirement for a definite complex between the molecules being separated and the chiral solvent molecules, in which at least two hydrogen bonds are present, to give a degree of conformational rigidity to the complex. Merely surrounding the enantiomers with chiral solvent molecules is insufficient, as is the formation of a single H bond. Striking success was achieved by Cram and coworkers (Cram and Cram, 1978; Kyba *et al.*, 1973) by the use of synthetic "host" molecules having a rigid structure and several H-bonding sites directed to the interior of a cavity, into which one "guest" enantiomer with corresponding outward-directed H-bond sites fitted, and the other enantiomer did not. Peacock and Cram (1976) reported the resolution of phenylglycine with 96% enantiomer purity in one extraction.

A reaction that in an achiral solvent would produce a racemic product, when carried out in a chiral solvent may result in the predominance of one of the enantiomers. This may result either from differential solvation of the reactants or of the transition state. Bosnich (1967) has shown that a symmetric solute in a chiral solvent may exhibit induced asymmetry, which can influence the ratio of enantiomers formed. Where the effect is through the transition state, one would expect the effect to be greatest if the solvent were involved directly, especially (in view of the foregoing) if two or more H-bonds are involved. If A and B are non-chiral molecules, but their adduct AB exists as enantiomers AB+ and AB−, the activated complexes in the reactions;

$$A + B + S + \rightarrow [SAB + +]^{\ddagger} \rightarrow S + + AB +$$

and                                                                                    (7.8)

$$A + B + S + \rightarrow [SAB - +]^{\ddagger} \rightarrow S + + AB -$$

are diastereomers, so the reactions would be expected to proceed at different rates, and yield more of one product than the other. If the reactions are reversible, and proceed to equilibrium, the racemic mixture will result, unless the solvent forms diastereoisomeric complexes with the products, in which case some enantioselectivity may still be obtained.

Rau (1983), in a review of asymmetric photochemistry in solution, briefly describes some effects of chiral solvents in photochemical synthesis. He mentions optical purities of products ranging from less than 1 to 23.5%. Reichardt (2003, p. 69) cites similar figures for a number of reactions, concluding that, while chiral solvents can induce asymmetry (or enhance the purity of products of reaction between chiral reactants) the effects are usually rather small. Baudequin *et al.* (2005) describe the new class of *chiral ionic liquids* (CILs). They present a long list of examples, with either the cation or the anion being chiral, including examples "tailor-made" for particular applications. They discuss applications to synthesis, to polymerization, to gas chromatography, and to NMR, and describe chiral liquid crystals. They suggest that CILs have great promise, and represent a "second wind" for chiral solvents. Bika and Gaertner (2008) review applications of chiral ionic solvents.

# 8

# NEW SOLVENTS AND GREEN CHEMISTRY

## 8.1 NEOTERIC SOLVENTS

A number of classes of solvents have been called "neoteric" (novel, newfangled). These include the room-temperature ionic liquids, "fluorous" solvents (perfluorinated or with perfluorinated tails), and supercritical fluids, chiefly water and carbon dioxide, though other examples such as the lower alcohols have been used. Recent work has highlighted the potential of biologically based solvents for their special properties and "green" character (see, e.g., Gu and Jérome, 2013). These new solvents offer a range of properties that have raised hopes of better synthetic methods and major improvements in environmental and workplace safety.

## 8.2 SUPERCRITICAL FLUIDS

For any substance in the supercritical state, that is, at a temperature above its critical temperature, the distinction between liquid and vapor (gas) is not relevant. The properties of a supercritical fluid, especially its solvent power, can be varied over a considerable range by adjusting the pressure. A number of substances whose critical temperatures or pressures are not inconveniently high have been used as solvents in this way. Typically, a desired product of a reaction, or a substance to be obtained from a natural source material, is extracted by the fluid at high pressure, and then caused to precipitate simply by lowering the pressure. The solvent is recompressed, to reenter the cycle. Sahu (2003) lists 13 substances that are in use as supercritical solvents. They include carbon dioxide, water, ammonia, and a number of lower hydrocarbons and chlorofluorocarbons. Water, at 647.3 K and 221.2 bar, has both the highest critical temperature and the highest critical pressure in the list, but is still in frequent use.

*Solvent Effects in Chemistry*, Second Edition. Erwin Buncel and Robert A. Stairs.
© 2016 John Wiley & Sons, Inc. Published 2016 by John Wiley & Sons, Inc.

### 8.2.1 Carbon Dioxide

"Naturally decaffeinated" coffee is made by extraction before roasting, using supercritical carbon dioxide ($scCO_2$)—an odd use of the word "naturally." $CO_2$ is nondipolar, but has a significant quadrupole moment. It is a weak HBA and a weak Lewis acid (presumably by accepting electrons into a $\pi^*$ orbital). In the supercritical states (plural because its density can be greatly varied by varying pressure) it is completely miscible with gases such as $H_2$, CO, and $O_2$. The power of $scCO_2$ as a solvent for most low to middling molecular mass organic substances and for water can be adjusted by varying its density in this way. It is useful in catalytic hydrogenation, as concentrations of $H_2$ much greater than are possible in common solvents may be achieved, but it can itself be catalytically reduced to CO, formic acid, or formaldehyde. It is resistant to oxidation, so it is a good solvent for reactions involving oxidation, particularly with $O_2$. The solvent properties of certain conventional solvents can be adjusted by "gas expansion," that is to say, by dilution with $CO_2$ at temperatures above or slightly below the critical temperature of pure $CO_2$, 31.1°C. In most applications of $scCO_2$ the products may be recovered by lowering the pressure. The $CO_2$ can then be vented or recovered by recompression.

Sahu (2003) briefly reviews uses of $scCO_2$ in cleaning and degreasing manufactured objects and in the foodstuffs, pharmaceutical, and plastics industries. In drug manufacture, $CO_2$ is effective in preparing active ingredients that are too soft to be mechanically ground in suitably fine crystalline form in two ways. As $scCO_2$ it can dissolve a drug at high pressure, and precipitate it on rapid decompression. Alternatively, a drug dissolved in a conventional solvent can be precipitated by introducing $CO_2$ at ordinary temperature but high pressure, when it acts as an antisolvent.

### 8.2.2 Water

Supercritical water is significantly less polar than water at ordinary temperatures and pressures. Consequently it dissolves many organic substances that have no significant water solubility at room temperature. This fact is exploited in the "SCWO" process for, among other things, destroying hazardous substances. In supercritical water at elevated temperature oxygen attacks such substances as chlorine-bearing or phosphorus-containing pesticides or chemical warfare agents, giving only $CO_2$, water, HCl, $H_3PO_4$, and so on, as relatively harmless products. A virtue of the process is that as the reactions are exothermic; once the desired temperature is reached, little or no further heat is required. It is not of as wide applicability as was first claimed, as there are problems, including corrosion of the containment vessel (supercritical water alone is somewhat aggressive, and any acid produced makes matters worse) (Kritzer and Dinjus, 2001). It is best used for hazardous substances not otherwise easy to dispose of.

The increased reactivity of water at the high critical temperature (374.7°C) and pressure (220.6 bar) can be a two-edged sword. The degree of hydration of both $H^+$ and $HO^-$ ions is reduced, so their activities are increased. (This is comparable, for

HO⁻, to the enhancement of its activity in aqueous dimethyl sulfoxide as the concentration of water approaches low values.) Wang *et al.* (2011) demonstrated methylation of phenol and benzene-1,2-diol derivatives with 1,3,5-trioxane (e.g., Scheme A) in supercritical water, accelerated compared to the solvent-free reaction. The authors describe the reaction as "uncatalyzed," though acid catalysis by water or by hydrogen ion is likely (see Section 3.6.1).

**SCHEME A**

Free-radical reactions are also promoted. Fujii *et al.* (2011) show that oxidation of methanol by $O_2$ in supercritical water involves both $HOO^•$ and $HO^•$ in a chain reaction.

## 8.3   IONIC LIQUIDS

Until fairly recently, "ionic liquids" meant "fused salts," and referred to substances with melting points well above room temperature (see, e.g., Bloom and Hastie, 1965; Kerridge, 1978). By using binary or ternary eutectics, lower working temperatures were attainable (e.g., the ternary eutectic of Li, Na, and K nitrates, 125°C), but these systems attracted little interest from organic chemists. Inorganic chemists, however, have found molten salts or salt mixtures useful for preparation of species otherwise unavailable. For instance, in liquid LiCl, the concentration of $Cl^-$ is high enough to cause the formation of species such as $FeCl_4^{2-}$ or $CrCl_6^{3-}$. In the curious room-temperature liquid, triethylammonium dichlorocuprate(I), even $FeCl_6^{4-}$ can exist (Porterfield and Yoke, 1976). The alkali metal halides dissolve notable amounts of their parent metals, giving solutions that resemble the metal–ammonia solutions in their optical, electrical, and magnetic properties. Some other metals may dissolve in their halide melts with reaction. For instance, while a solution of cadmium in fused cadmium chloride gives back the metal on cooling and solidifying, addition of $AlCl_3$ causes the precipitation of $Cd_2^{2+}[AlCl_4^-]_2$, containing the dimeric cadmium(I) ion, analogous to mercurous ion, $Hg_2^{2+}$.

Many studies have been related to extractive metallurgy or to investigation of reactions responsible for the formation of minerals, including ores. The chemistry of molten oxide mixtures led to the formulation by Lux (1939) of the acid–base theory, refined by Flood *et al.* (1952), based on oxide ion as the species transferred, analogous to the proton in the Brønsted–Lowry view, though opposite in sense, for an oxide donor is a base, and an oxide receptor is an acid. Many authors in the past put "acid" and "base" in quotation marks, or used such phrases as "acid analogue" in discussing the chemistry of such melts. This is quite unnecessary, however, as the analogy with protonic acidity is perfect, *mutatis mutandis*. Highly silicic lavas, slags, and so on, are acidic, and rocks with less than about 50% $SiO_2$ can be called "basic" without apology, as demonstrated by measurements of pO (negative common logarithm of the $O^{2-}$ activity) using suitable electrodes (El Hosary *et al.*, 1981).

Ethylammonium nitrate, m.p. 14°C, was described by Walden in 1914, but did not attract wide attention. The discovery of the 2:1 aluminium trichloride/ethylpyridinium bromide combination, which is also liquid at room temperature (Hurley and Weir, 1951), began a new chapter, introducing a family of systems involving $AlCl_3$ and halides of large organic cations. Acid–base properties of $AlCl_3/MCl$ systems ($M^+$ alkali metal cation) had previously been studied, using an Al wire electrode to measure $Cl^-$ activity. The neutral species (in the acid/base sense) is the $AlCl_4^-$ anion. The reaction (called by Chum and Osteryoung (1981) *autosolvolysis*):

$$2AlCl_4^- \rightleftharpoons Al_2Cl_7^- + Cl^-; \quad K_m \tag{8.1}$$

is analogous to autoprotolysis in water. $Al_2Cl_7^-$ is the acid species and $Cl^-$ the basic species—an unusual pair as both are anions.

The equilibrium constant for this reaction is strongly temperature dependent and somewhat influenced by the nature of the cation. Chum and Osteryoung report values of $K_m$ of $1.06 \times 10^{-7}$ in $AlCl_3/NaCl$, $<1.19 \times 10^{-8}$ in $AlCl_3/N$-$n$-butylpyridinium chloride (RCl) at 175°C, and $3.8 \times 10^{-13}$ in the latter system at 30°C. They show what are effectively titration curves, showing a potential jump of more than $-0.8\,V$ as the mole ratio of $AlCl_3/RCl$ passes from just below to just above unity, corresponding to a change in pCl at $30°C$ of nearly 14 units, fully comparable to the pH jump in the titration of aqueous $1\,M\,NaOH$ with strong acid.

Anions other than the very water-sensitive and reactive haloaluminates have been added to the repertoire, including nitrate, phosphate, hexafluorophosphate, hexafluoroanti-monate, tetrafluoroborate, trifluoromethanesulfonate ("triflate"), and bis (trifluoro-methanesulfonyl)imide, $(CF_3SO_2)_2N^-$. Examples of cations that are used include mono- to tetra-alkylammonium and phosphonium, 1-alkyl-3-methylimidazolium, **1**, N-alkylpyridinium, **8** and the ions **3, 4, 5**, and **6** containing more than one quaternary nitrogen atom. A wide variety of properties is thus available, including water solubility from near zero to total miscibility, and selective solubility of different classes of organic substances. Plešek and Heřmánek (1968), in discussing the uses of sodium hydride, comment on its "complete insolubility in the usual solvents," though it is soluble in molten sodium hydroxide. One wonders if there are any room-temperature ionic liquids that would dissolve sodium hydride without reacting with the exceedingly strongly basic hydride ion.

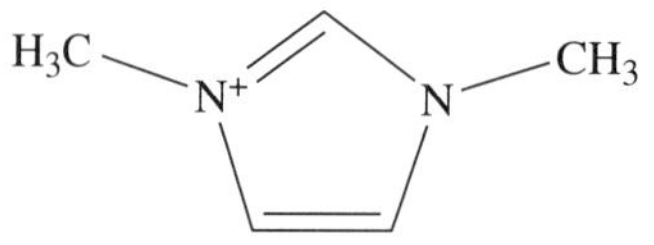

**1** 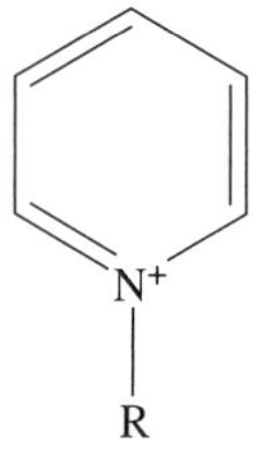 **2**

Room-temperature ionic liquids have recently attracted a great deal of industrial interest (Carmichael, 2000; Guterman, 1999; see also the brief overview by Earle *et al.* (2003) and the extensive and detailed review by Olivier-Bourbigou *et al.* (2010). They are versatile as solvents or nonsolvents for organic substances, and some exhibit strongly temperature-dependent water solubility. They are nonvolatile. Some, notably those containing haloaluminate anions, have widely variable Lewis and Brønsted acidity (into the superacid range). Others, such as those with tetrafluoroborate or hexafluorophosphate

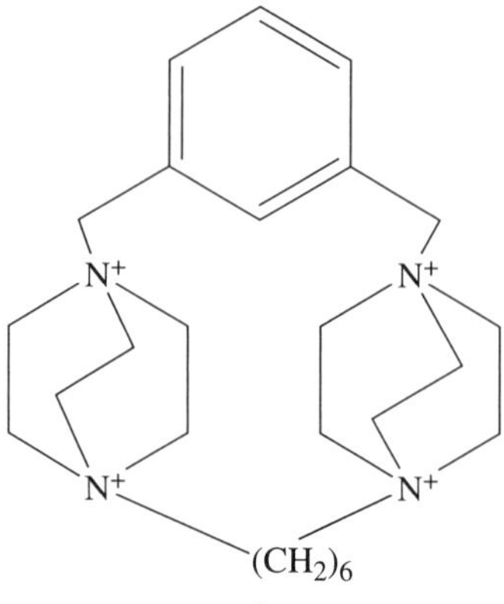

**3**

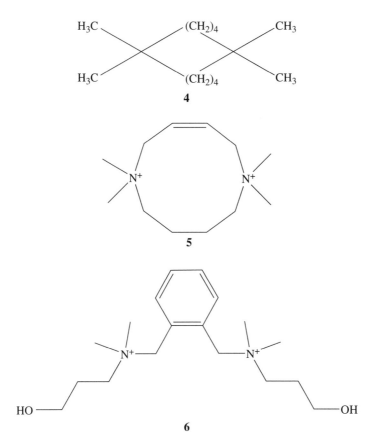

anions, lack strong acidic or basic properties. Those with anions like $CF_3SO_3^-$ or $(CF_3SO_2)_2N^-$ take up less water when exposed to moist air (Endres, 2002). These unreactive examples have the ability to dissolve metal-containing catalysts without deactivating them, permitting recovery and recycling of catalysts that would otherwise have been lost to waste. By adjustment of the temperature, certain reaction mixtures can be made homogeneous for reaction, and then cooled to cause phase separation for recovery of products; or the reaction mixture can be maintained as a two-phase system while the reaction proceeds.

### 8.3.1 Reactions in Ionic Liquids

Ionic liquids, owing to their stability and nonvolatility, contribute to "green" chemistry. In many applications they are easily recovered and recycled, and volatile organic substances can be recovered from them by distillation under mild conditions *in vacuo*. For instance, the ability of 1-butyl-3-methylimidazolium hexafluorophosphate to dissolve both the organic reagent (chloromethyl)-benzene and the salt potassium cyanide has been exploited (Wheeler *et al.*, 2001) in a successful preparation of phenylacetonitrilc, with minimal environmental impact. Headley and Ni (2007) describe the synthesis and uses of chiral ionic liquids, and suggest that they may open a new route to asymmetric synthesis.

The use of ionic liquids in synthesis is the subject of a monograph by Wasserscheid and Welton (2008). A number of reaction types may be advantageously carried out in room-temperature ionic liquids. They include the following:

### 8.3.1.1  Friedel–Crafts

*8.3.1.1  Friedel–Crafts*   The variable acidity afforded by chloroaluminate ionic liquids enables improved control of the course of Friedel–Crafts acylations and alkylations over traditional solvents. Boon *et al.* (1986) showed that the rate of acetylation of benzene (Eq. 8.2) by acetyl chloride in chloroaluminate liquids was directly proportional to the concentration of $Al_2Cl_7^-$. That the steric course of reaction may be controlled is illustrated by the predominance of α- over β-acetylation of naphthalene in 1-ethyl-3-methylimidazolium chloride/AlCl$_3$ (Adams *et al.*, 1998); the reverse is true in nitrobenzene as solvent. Alkylation of benzene with 1-dodecene, using $[(CH_3)_3NH]^+ \cdot Al_2Cl_7^-$ as catalyst gives a better yield of linear dodecylbenzene for detergent manufacture, with less loss of catalyst, than the similar process using liquid HF (P. de Jonge, cited by Carmichael, 2000). A variant method uses a neutral ionic liquid, and scandium(III) triflate as an immobilized Lewis acid catalyst (Song *et al.*, 2000).

$$\text{(8.2)}$$

### 8.3.1.2  Diels–Alder

*8.3.1.2  Diels–Alder*   The activity as dienophiles of unsaturated esters, or aldehydes such as crotonaldehyde (2-butenal), toward cyclopentadiene in dichloromethane was enhanced by the addition of 0.2 equivalent of 1,3-diethylimidazolium bromide or acetate (Howarth *et al.*, 1997). Diels–Alder reactions have also been carried out in ionic liquids as the sole solvents, with or without the addition of Lewis acid catalysts (Song *et al.*, 2001). The yields were generally satisfactory, and exo/endo selectivity moderately strong (in the region 4.3–4.9 to 1), comparable to ratios found in polar solvents. Notably, for Reaction 8.3 in 1-ethyl-3-methylimidiazolium chloride/ aluminum chloride solvents as the solvent composition was changed from the basic to the acidic region, the selectivity changed from 4.88: 1 to 19: 1, with some loss of yield. Kumar (2001) comments that it is too soon to assess or explain the effects of these ionic media on Diels–Alder reactions, and that more kinetic data are required. He refers to the review by Welton (1999) for a discussion of the properties of ionic liquids that are relevant to their usefulness in this context, and in organic chemistry generally.

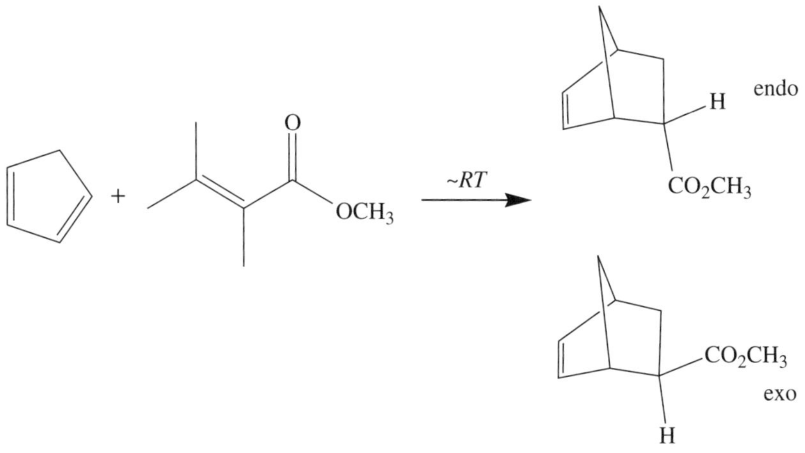

(8.3)

### 8.3.1.3 Hydrogenation and Hydroformylation

*8.3.1.3 Hydrogenation and Hydroformylation* Platinum group metals, in the form of low oxidation state complexes or cluster compounds dissolved in ionic liquids such as 1-butyl-3-methylimidazolium tetrafluoroborate, have been used as hydrogenation catalysts for both aliphatic (Suarez *et al.*, 1997) and aromatic (Dyson *et al.*, 1999) substrates. The reactions proceed heterogeneously, and the product is easily separated. The catalyst remains in the solvent for reuse. This method has been used successfully to hydrogenate acrylonitrile–butadiene rubber, rendering it less susceptible to thermal and oxidative degradation (Suarez *et al.*, 1997). Hydroformylation reactions (e.g., Reaction 8.4) in ionic liquids afford yields similar to those obtained in dichloromethane as solvent, but separation of the product is easier, and loss of a volatile organic substance to the environment is minimized.

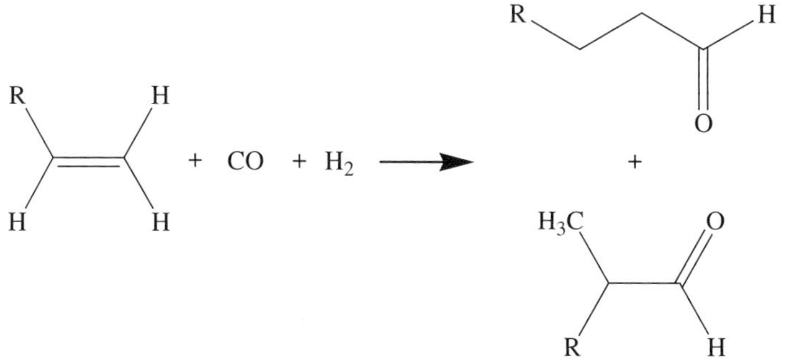

(8.4)

### 8.3.1.4 Oligomerization

*8.3.1.4 Oligomerization* Dimerization or oligomerization of low molecular mass alkenes is regularly carried out industrially, with the aim of preparing linear 1-alkenes for synthetic use. In these applications, the presence of strong Lewis acids is undesirable, so neutral ionic liquids are used, exemplified by [bmim]$PF_6$. ([bmim]$^+$ is 1-butyl-3-methylimidazolium.) This has been successful in the manufacture of hexenes from ethylene using a nickel complex as catalyst (Wasserscheid *et al.*, 2001), with conversion of 95%, the product being 92% 1-hexene. An earlier study by Chauvin and coworkers (1995) on the dimerization of propene demonstrated the pitfalls of using chloroaluminate liquids: in the acid region, cationic side reactions led to too great molecular masses and high viscosity, while in the basic region, chloride displaced ligands from the nickel catalyst, deactivating it. It was shown that the system may be buffered by adding an insoluble solid alkali metal chloride to the liquid on the acidic side (Ellis *et al.*, 1999), which converts excess $Al_2Cl_7^-$ into neutral $AlCl_4^-$ and at the same time provides alkali cations to precipitate chloride, preventing an excursion into the basic region. Alternatively, replacing $AlCl_3$ with $EtAlCl_2$, which is less acidic, led to successful production of the desired dimers.

An avenue that is now being explored involves the use of "task specific" ionic liquids, in which one of the ions is furnished with a functional group for a particular purpose. An example shown is Reaction 8.5 (not balanced), developed by Davis and coworkers (Bates *et al.*, 2002). The amino group binds carbon dioxide as a carbamate at room temperature in a reaction that is reversed at a higher temperature. It is proposed as a substitute for aqueous amines now used to scrub carbon dioxide from natural gas. The new process avoids problems with loss of the amine and of water. A problem with viscosity was encountered, which may perhaps be overcome by changing the butyl side chain, or by suitable dilution. This is an example of a type of process that should become important in many industries, for extraction of toxic or valuable substances from gaseous or liquid waste streams. Ionic liquids can be prepared that are essentially immiscible with water. A suitable ionic liquid denser than water coming down a packed tower could extract many organic or inorganic solutes from an ascending aqueous stream. Losses of the ionic solvent to the water are expected to be very low, and the solvent and the extracted substances should be recoverable by moderate heating under reduced pressure.

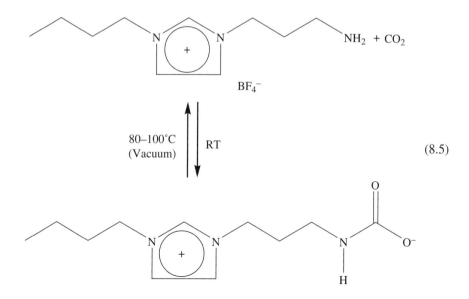

(8.5)

Reichardt (2003, pp. 59–62, 322–324) mentions a number of applications of ionic liquids, and cites several reviews. Zhang *et al.* (2006) have compiled a list of over 600 ionic liquids and their properties. Baudequin *et al.* (2005) present a long list of chiral ionic liquids and discuss their uses (see earlier, Section 7.4). Endres (2002) and Ohno (2011) review the electrochemical possibilities of these liquids, which are many and varied. See also Bika and Gaertner (2008), Headley and Ni (2007).

A recent review by Tang *et al.* (2012) describes methods of preparation of ionic liquids in which either the cation or the anion beas functional groups such as hydroxyl, ethers (including PEG: $-(CH_2-CH_2-O)_n-CH_3$ with *n* up to 16), thiols, and others. They discuss their physical, including electrochemical, properties, toxicity and biodegradability, and applications as solvents for organic and enzymatic reactions, as media for capture of $CO_2$ and $SO_2$ and for separation of gaseous mixtures. They list 406 references to literature.

## 8.4 LOW-TRANSITION-TEMPERATURE MIXTURES

Tang and coworkers (2012) mention a related class of liquids, so-called deep eutectics (DEs), otherwise known as "low-transition- temperature mixtures," described by Abbott *et al.* (2003, 2004) and by Francisco *et al.* (2013). These are prepared by simply warming together any of a large number of hydrogen-bond donors with any of a group of acceptors. Examples the latter workers describe include urea + choline chloride, malic acid + alanine. These liquids share some properties with ionic liquids, and are similarly versatile, as the choice of each of the two components is wide. Solubilities of an unusual variety of substances in certain of these mixtures are reported, including both organic and inorganic substances that have low solubilities in common solvents. Some of the preparations showed no freezing point, but on cooling increased in viscosity until they became glassy. (This is why Francisco *et al.* prefer the term "low-transition- temperature mixtures.")

Abbott *et al.* (2003) describe, as a typical example, mixtures of urea (m.p. 132.7°C) with choline chloride (m.p. ~310°C), which exhibit a eutectic at *ca.* 64 mol percent urea and 12°C. The fluidity (reciprocal of viscosity) and electrical conductivity of the eutectic mixture increase together with increasing temperature. The latter is similar to that of many ionic liquids, approximately $1\,mS\,cm^{-1}$ at 30°C. They report solubilities at 50°C, in mol/l: LiCl >2.5; AgCl 0.66; benzoic acid 0.82; D-alanine 0.38; CuO 0.12.

Francisco *et al.* (2013) describe further the properties of these mixtures as solvents and reaction media. They note that their preparation is in most cases much simpler and less ecologically undesirable than ionic liquids, as they are prepared by simply mixing the two components. No changes of covalent bonds are required. They present charts showing the fields in which they have been applied, including electrochemistry, the preparation of novel materials, synthesis, and separation processes. The chicf drawback of DESs they mention is that they are too water-soluble to be used in two-phase systems with water, for instance, extraction from aqueous solutions. Their thermal stability at higher temperatures is yet to be studied.

## 8.5  BIO-BASED SOLVENTS

Petroleum has been a major source of solvents hitherto, but a more sustainable source would be desirable. Plant-derived substances are naturally biodegradable. Work is now going forward in the search for liquids from renewable sources that can replace petroleum-derived hydrocarbons or derivatives. Gu and Jérome (2013) list 146 references to this growing field. (Pun intended!) Examples of polar solvents described include glycerol and derivatives, which figure in 44 of the papers cited, carbohydrates, in aqueous solution or in low-melting mixtures, lactic acid and its esters, and aqueous gluconic acid, all of which offer advantages as media for certain reactions. Less polar examples include 2-methyltetrahydrofuran and γ-valerolactone, fatty acid methyl esters, and as a replacement for petroleum distillates, D-limonene, obtained by distillation of waste citrus peel. Lignin, a waste in paper-making, is cited as a possible source of a number of special solvents.

Some of these efforts to find useful new, ecologically friendly solvents complement each other. Choline chloride forms low-melting mixtures with carbohydrates (Rusz and König, 2012) and lowers the viscosity of glycerol.

## 8.6  FLUOROUS SOLVENTS

Fluorous solvents include those fully fluorinated, such as perfluoroalkanes (*f*-alkanes), and molecules with a perfluorinated "tag," such as enneafluorobutyl methyl ether, $C_4F_9OCH_3$. Their usefulness springs from their immiscibility with both polar solvents (including water) and nonpolar, ordinary organic solvents at room temperature, and miscibility at higher temperature. Horváth and Rábai (1994) demonstrated this by carrying out the hydroformylation of 1-decene in mixed toluene and *f*-methylcyclohexane, with a fluorous rhodium complex in which one of the ligands was tris-(2-*f*-butyl-ethyl) phosphane as a catalyst, at 100°C. On cooling to room temperature, the homogeneous phase separated into a fluorous phase containing the catalyst and an organic phase containing the product, the desired linear undecanal in 85% yield. A report of Fluorous Technologies Ltd. (2011) illustrates a three-phase system, in which an organic layer, an aqueous layer, and a fluorous layer are cleanly separated. A reaction in which inorganic, organic, and fluorous substances are involved, whether as reactants, catalysts, or products, can be carried out at a high temperature in a mixture of a fluorous solvent, water, and an organic solvent. On cooling, the fluorous solutes go into the fluorous phase, inorganics into the water layer, and the organic products into the organic layer. Fluorous solvents may be "tuned" by changing the size or number of the fluorous substituents, to be more or less miscible with organic liquids.

## 8.7  SWITCHABLE SOLVENTS

A switchable solvent is one that can be made to change important physical properties, such as polarity or hydrophobicity, under the influence of an external factor, which may be a change of temperature or pressure or the addition or removal of a gas, such as carbon dioxide. Pollett *et al.* (2011) in a "mini-review" provide references to a number of studies, and discuss two examples. Both involve gaseous cosolvents. Piperylene sulfone (**1**) has been proposed as an alternative to dimethyl sulfoxide (Vinci *et al.*, 2007). It can be prepared from $SO_2$ and commercially available piperylene (1,3-pentadiene, **2**) in the presence of a radical inhibitor (to prevent polymerization of the latter). Only the *trans* isomer reacts at an effective rate. The product may be used as a solvent for a reaction. On completion, raising the temperature to near 100°C causes decomposition of the solvent to $SO_2$ (b.p. 10°C) and pure *trans*-piperylene (b.p. 42°C), which may be cocondensed and allowed to react to reform the sulfone. The products of reaction, provided they are relatively high boiling, may be recovered from the now solvent-free residue. Vinci *et al.* (2007) carried out several successful syntheses involving nucleophilic substitutions on (chloromethyl)benzene. They found reaction rates comparable to, though somewhat slower than, similar reactions in dimethyl sulfoxide. They attribute solvent losses mainly to the small scale of their preparations, and suggest that the losses would be much less in industrial-scale operations.

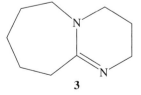

The other example that Pollett *et al.* describe is a class of molecular liquids that can be converted to ionic liquids by reaction with $CO_2$. The reactions are of the general form of Equations 8.6 (one component) or 8.7 (two components).

$$2RNH_2 + CO_2 \rightleftharpoons RNH_3^+ + RNHCOO^- \tag{8.6}$$

$$RNH_2 + R'OH + CO_2 \rightleftharpoons RNH_3^+ + R'OCOO^- \tag{8.7}$$

The R groups need to be fairly bulky, for example, R = 3-trialkylsilylpropyl or 3-trialkoxysilylpropyl in one example cited. The nitrogenous compound in the two-component example may by heterocyclic, for example, 1,8-diazabicyclo[5.4.0]undec-7-ene (DBU, **3**).

**3**

They describe a Heck reaction of bromobenzene with styrene to give stilbene, carried out in the ionic form of DBU + hexanol + $CO_2$. The product *E*-stilbene (in 97% yield) was extracted into heptane. Upon removal of $CO_2$ the byproduct salt DBU.HBr precipitated from the now molecular solvent mixture. The catalyst, which was a palladium complex, remained in solution.

An example applicable to the food industry is described by Jessop *et al.* (2010). Soybean oil is produced industrially by extracting the crushed beans with hexane (a known neurotoxin), which is then removed by distillation. A number of amidines and guanidines were tested with the aim of finding a suitable switchable hydrophilicity solvent (SHS), one which could be converted from a hydrophobic to a hydrophilic form by reaction with carbon dioxide. Some were too hydrophilic to start with, while others (guanidines) reacted so strongly with $CO_2$ that the reaction could not be easily reversed. One that was successful was *N,N,N'*-tributyl pentanamidine, **4**. When equal volumes of **4** and water were stirred together and exposed to $CO_2$ at 1 bar, they reacted according to Equation 8.2, and formed a homogeneous liquid phase. On removal of $CO_2$ by aeration, it reverted to two phases, essentially pure liquid **4** and water. Pure **4** was shown to be as efficient a solvent as hexane for extraction of oil from soybean flakes.

A process was proposed, illustrated in Figure 8.1. Oil is extracted from crushed beans with **4**. Addition of water and $CO_2$ to the extract causes formation of a solution of the hydrophilic salt and water; the oil is expelled as a separate phase, and separated. Removal of $CO_2$ by vacuum restores the hydrophobic liquid, **4**, and it, the water, and $CO_2$ are recycled.

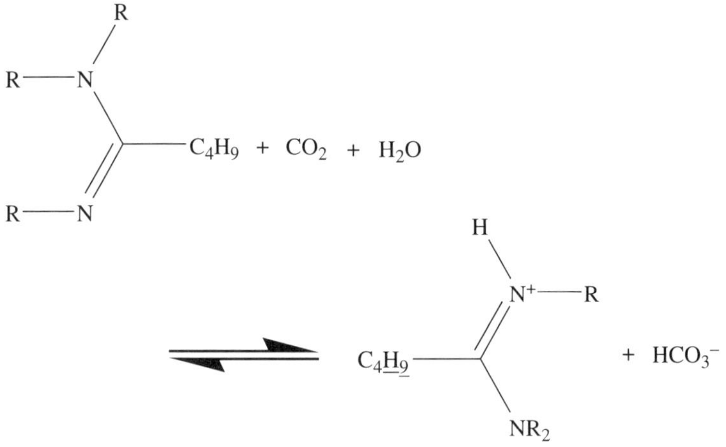

## 8.8   GREEN SOLVENT CHEMISTRY

"Greenness" as it relates to solvents has two aspects: the nature of the solvent and how it is used. Some solvents are obviously innocuous, water being the obvious example, though its high heat of vaporization makes its separation from reaction products by distillation a potential problem. (Partial freezing or reverse osmosis demands less energy.) Others are easy to obtain or manufacture and are environmentally benign, being biodegradable or otherwise nonpersistent. Ethanol comes to mind. Some are poor in one aspect, but good in the other. Examples are some of

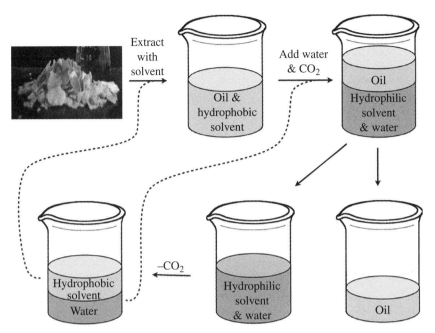

**FIGURE 8.1** The process by which a switchable-hydrophilicity solvent can be used to extract soybean oil from soybean flakes without a distillation step. The dashed lines indicate the recycling of the solvent and the aqueous phase. From Jessop *et al.* (2010) by permission of the Royal Society of Chemistry. (*See insert for color representation of the figure.*)

the ionic liquids, which are expensive and environmentally difficult to manufacture, but in certain uses are readily separated from reaction mixtures for recycling, and low-boiling hydrocarbons, which are easily obtained from petroleum in the course of refining but are easily lost through evaporation and are contributors to photochemical smog.

Before a solvent comes into use, whether in a laboratory, a factory, or a household, much has happened. The raw materials have been assembled, the process of manufacture and purification carried out, the product packaged and shipped. Each stage involves energy for transportation or processing and some degree of waste needing disposal. To be considered "green," in the sense that it is not harmful to the environment, a solvent cannot therefore be judged merely on its "downstream" effects, that is, those consequent on its use, recovery, and recycling.

Jessop (2011) points out that ideally a life cycle analysis (LCA) should be done, to assess the degree of environmental hazard arising from the whole history of the solvent, from raw materials to ultimate disposal, but acknowledges that this is rarely possible for the working chemist. In the absence of an LCA, he has outlined a protocol aimed at examining the manufacturing stage, based on a "synthesis tree," like a family tree, with the solvent at the bottom, its "parents" (the substances directly used in its preparation) next above, and so on upward until the original substances mined or otherwise obtained from the natural world are reached. The number of steps in such a tree will be very small for some substances: water, simple hydrocarbons, or ethanol. For some it may be large. Jessop shows one for the room-temperature ionic liquid [bmim]BF$_4$, after Zhang *et al.* (2008), of 30 steps. Each step may be examined to see whether it poses a workplace or environmental hazard (e.g., HF, explosion) or a large energy requirement (e.g., distillation).

A method of assessing the downstream effects of emission of solvents to the environment is described by Gama *et al.* (2012). It is based on the RAIDAR model (Arnot and Mackay, 2008; Mackay *et al.*, 1992). This mass-balance model uses the concept of fugacity to predict the behavior of a substance emitted at a constant rate into air, water or soil, or some combination of these three receiving environments. For each substance the factors considered include its hydrophobicity (as measured by $K_{ow}$), its chemical reactivity (susceptibility to hydrolytic, oxidative, microbial, or photochemical degradation), and its toxicity. Toxicity assessment includes a baseline due to nonspecific narcosis, a common characteristic of all nonpolar substances owing to their tendency to accumulate in lipids and interfere with membrane function. A toxic ratio (TR) is used to assess the potential for toxic effects for a substance that exhibits a specific mechanism of toxicity, that is, that exerts effects at concentrations lower than baseline narcosis. Gama *et al.* (2012) show results, for 22 commonly used solvents, of applying the model in an "environment" made up of an area of 100,000 km$^2$ with 10% fresh water with sediment and 90% soil, under 1,000 m of atmosphere, and representative ecological receptors (biota). Tables show input values and assumptions used for the model, and the resulting persistence estimates (as residence times) and long-range transport, as characteristic travel distance (CTD), and more detailed results for four of the chosen solvents, including predictions of what classes of organisms are most vulnerable to each. In these examples, photolysis and hydrolysis were explicitly neglected, and the unrealistically low emission rate of 1 kg h$^{-1}$ was assumed. The authors point out that, as the equations used in the model are all linear in the concentrations, all the results can be scaled up, in proportion to more realistic emission rates. A figure reprinted from Arnot and Mackay (2008) shows the ways that one example, perchlorethylene, moves among and is distributed to the four environmental subspheres: air, soil, water, and sediment.

Reichardt and Welton (2011, p. 510) cite Anastas and Warner's (1998) 12 Principles of Green Chemistry, which we present in abbreviated (and imperative) form:

1. Prevent waste rather than clean up afterward.
2. Design synthesis to incorporate most of the inputs in the product.
3. Avoid use or generation of harmful substances.
4. Design products to maintain usefulness while minimizing toxicity.
5. Minimize solvent use, or use innocuous ones.
6. Minimize energy use.
7. Use renewable feedstocks.
8. Avoid unnecessary derivatization.

9. Prefer catalysts to stoichiometric reagents.
10. Design products to be nonpersistent in the environment.
11. Monitor and control processes in real time.
12. Minimize risk of accident: releases, explosions, fires.

For the undergraduate teaching laboratory, many add a 13th: Use microscale experiments.

In synthesis, the fewer steps, and the more steps that can be carried out without change of solvent, the better. Reichardt and Welton (2011, p. 512) mention two major successes in reduction of solvent use by Pfizer: an amazing 180-fold reduction in solvent use in the synthesis of Viagra™ from discovery to the final, optimized scheme (Dunn et al., 2004), and the three steps in different solvents with isolation of intermediates at each stage in the synthesis of sertraline done all in ethanol with isolation of the final product only (Taber et al., 2004).

Reichardt (2003, p. 504) presents a table of solvents recommended as replacements for more hazardous ones. Total avoidance of solvent is sometimes a possibility. An example was mentioned in Chapter 1. Scott (2003) discusses the scope of solvent-free methods and considerations of mutual solubility of reactants and of control, including exothermic reactions. A recent application is the solvent-free synthesis by Valdez-Rojas et al. (2012) of thioamides and alpha-ketothioamides through the Willgerodt-Kindler reaction, Scheme B.

Capello et al. (2007) present the results of assessment of 26 common industrially used organic solvents, using two tools. One is an Excel-based, simplified Environmental Health and Safety procedure using the "EHS Tool" (Sugiyama et al., 2006) that looks at the available data for a substance, physical and chemical properties, and toxicity, to assess its degree of hazard in the workplace and its probable environmental effects both locally and globally over time. Categories of hazards are in air, in water, persistency, chronic toxicity, acute toxicity, irritation, reactivity/decomposition, fire/explosion, and release potential. A point is assigned for each category considered severe, and downward to 0.1 point for nearly no hazard. The score for a typical solvent may be from approximately 2.6 (ethanol) to 5 (1,4-dioxane) or more.

The other uses the "Ecosolvent" tool (Capello et al., 2006) to consider the cumulative energy demand (CED) over a substance's whole life cycle. Three figures are arrived at. The first is the CED for its production, reported in megajoules per kilogram of solvent. The net CED after recovery by distillation is calculated assuming 90% recovery, as 10% of the CED for production plus the energy consumed by the distillation of the other 90%. The net CED after incineration is that for production less the energy that may be recovered by efficient incineration. A few examples are listed in Table 8.1.

A group of workers at Pfizer Global Research and Development (Alfonsi et al., 2008) have used these principles to draw up a list of solvents in a "traffic-light" scheme (green = preferred, amber = usable with caution, red = undesirable), together with a list of substitutes for undesirable solvents. They were considering the pharmaceutical industry primarily, but their advice is of general applicability. The

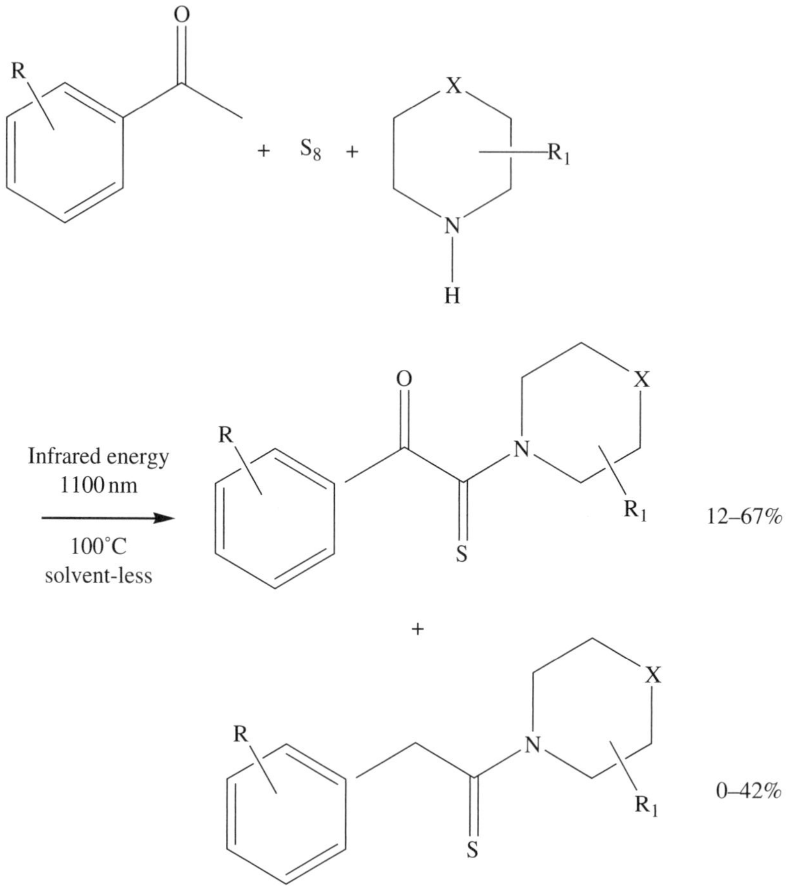

**SCHEME B**

**TABLE 8.1    Cumulative Energy Demand (CED) for Three Common Solvents, in MJ Kg$^{-1}$)$^a$**

| Examples | CED for production only | Net after recovery | Net after incineration |
|---|---|---|---|
| Toluene | 80.0 | 20.0 | 30.7 |
| Ethanol | 50.1 | 18.9 | 18.4 |
| 1-Propanol | 111.7 | 24.4 | 75.0 |

$^a$ From larger selection in Capello *et al.* (2006).

"undesirable" list contains one or two substances that are less bad than others. For instance, dichloromethane is listed, but yet it is offered as a substitute for the carcinogens trichloromethane, tetrachloromethane, and 1,2-dichloroethane. Their "traffic-light" list is shown here as Table 8.2. It does not include any of the neoteric solvents.

**TABLE 8.2 The Pfizer Solvent Selection Guide for Medicinal Chemistry**[a]

| Preferred (green) | Usable (amber) | Undesirable (red) |
|---|---|---|
| Water | Cyclohexane | Pentane |
| Acetone | Heptane | Hexane(s) |
| Ethanol | Toluene | Di-isopropyl ether |
| 2-Propanol | Methylcyclohexane | Diethyl ether |
| 1-Propanol | *t*-Butyl methyl ether | Dichloromethane |
| Ethyl acetate | Isooctane | 1,2-Dichloroethane |
| Isopropyl acetate | Acetonitrile | Trichloromethane |
| Methanol | 2-Methyltetrahydrofuran | *N,N*-Dimethylformamide |
| 2-Butanone | Tetrahydrofuran | *N*-Methylpyrrolidin-2-one |
| 1-Butanol | Xylene(s) | Pyridine |
| 2,2-Dimethylpropan-1-ol | Dimethyl sulfoxide | *N,N*-Dimethylacetamide |
| (*t*-Butanol) | Acetic acid | 1,4-Dioxane |
| | Ethane-1,2-diol (Glycol) | 1,2-Dimethoxyethane |
| | | Benzene |
| | | Tetrachloromethane |

[a] Reprinted from Alfonsi *et al.* (2008, p. 32) with permission of the Royal Society of Chemistry.

Probably most of them would appear in the "red" list. The ionic liquids are problematic owing to their complicated syntheses, the fluorous solvents because of the HF used in their manufacture, and the supercritical fluids because of the hazards of high pressure and, in the case of water, high temperature attendant on their use.

Finally, Jessop *et al.* (2010) issues four "grand challenges" to those working in the field of green chemistry:

1. Finding a sufficient range of green solvents. The green list in the Pfizer guide is the shortest.
2. Recognizing whether a solvent is actually green. All aspects need to be considered: its cumulative energy demand and environmental and workplace health and safety in manufacture, use, and ultimate disposal.
3. Finding an easily removable polar aprotic solvent. Existing examples require distillation; extraction with water merely postpones it.
4. Eliminating distillation.

# 9

---

# CONCLUDING OBSERVATIONS

## 9.1 CHOOSING A SOLVENT

Up to this point, we have been considering a number of aspects of solvents and their effects, party from a theoretical point of view, or as empirical observations. The reader may well say, "What now?" and expect some practical advice. What this advice will be must depend on the purpose of the procedure being undertaken.

Carlson *et al.* (1985) describe an approach to a formal method of choosing the most suitable solvent, based on a principal component analysis (see Chapter 4) of the properties of a set of solvents of diverse types. Using 82 solvents and 8 descriptors (m.p., b.p., relative permittivity $\varepsilon_r$, dipole moment $\mu$, refractive index $n_D^{20}$, $E_T(30)$, density $\rho$, and log $K_{o/w}$), they show that 51% of the variance of the solvent properties can be represented by two principal vectors (components), $t_1$ and $t_2$, and that a third does not add significance. (It may be observed that the descriptors chosen are all of the nonspecific type, with the exception of $E_T(30)$.) They then propose an experimental design based upon a systematic exploration of the surface defined by the vectors $t_1$ and $t_2$. The point of this formal procedure is to avoid premature choice of a single solvent type based upon an assumed mechanism, which, if it should prove unproductive, might lead the investigator to give up. They do not claim to have described a definitive procedure, but rather they point in a direction in which future work might proceed.

In the absence of a suitable formal method of choice, the choice must be guided by the information available and the purpose of the work. Is the solvent to be a reaction medium, either for synthesis or for a mechanistic study? Practically, it must have a reasonable liquid range, be reasonable easy to purify, be separable from reaction products (by evaporation or by their low solubility), and not react destructively with the reactants or hoped-for products. Considerations of workplace or environmental hazards should not be overlooked (see Sections 1.3 and 8.6, and Table A.1).

---

*Solvent Effects in Chemistry*, Second Edition. Erwin Buncel and Robert A. Stairs.
© 2016 John Wiley & Sons, Inc. Published 2016 by John Wiley & Sons, Inc.

Maximizing yield as described in Chapter 2 may or may not be important, though the effects of changing solvents may be decisive in assigning a mechanism. Certain reactions may require a two-phase system, if one reactant is ionic and the other relatively nonpolar. For instance, in the emulsion polymerization method of making certain synthetic rubbers (Billmeyer, 1971), a mixture of monomers (styrene and butadiene in the classic case, "GR-S," important during the Second World War), neat, is dispersed with the help of soap in an aqueous solution of the initiator, a source of radicals. Initiators used have included potassium persulfate and a reducing agent, and Fenton's reagent, which contains hydrogen peroxide and ferrous sulfate. The soap forms micelles containing monomer. HO radicals diffuse across the phase boundary and initiate radical polymerization within the micelle. Concentrations can be adjusted so that on average only one radical at a time enters each micelle, so each contains a single growing polymer.

A simplified scheme of reaction is as follows:

$$HOOH(aq) + Fe^{2+}(aq) \rightarrow HO^{\cdot}(aq) + HO^{-}(aq) + Fe^{3+}(aq) \quad \text{(initiation)}$$

$$HO^{\cdot}(aq) \rightarrow HO^{\cdot}(org) \quad \text{(phase transfer)}$$

$$\left.\begin{array}{l} OH\cdot + M \rightarrow HOM\cdot \\ \qquad \vdots \\ \qquad \vdots \\ HOM_i\cdot + M \rightarrow HOM_{i+1}\cdot \ (= R_{i+1}\cdot) \end{array}\right\} \quad \text{(propagation)}$$

$$R_i\cdot + R_j\cdot \rightarrow P_{i+j} \quad \text{(termination)}$$

M may be either monomer. $HOM_i\cdot$ or $R_i\cdot$ represents an active polymeric radical containing $i$ monomer units; $P_i$ represents a "dead" polymer. Chain-transfer reactions, in which a polymer radical abstracts an H atom from a monomer (starting a new chain) or from the middle of another polymer molecule (resulting in a branched structure), and other modes of termination are possible.

Some such two-phase reactions require a phase-transfer catalyst, usually a large organic ion, such as tetrabutylammonium, to carry an ionic reagent into the nonpolar phase.

Solvents for recrystallization must satisfy a fairly rigid set of criteria. The substance to be purified must be much more soluble at high than at low temperature. Expected impurities should be either very soluble or not at all. The solvent must be easily separated from the crystals by evaporation or soluble in a second solvent in which the crystals are not, so it may be washed off.

Solvents for chromatography, for electrochemistry, or for titrimetric analysis of weak acids or bases all must satisfy appropriate criteria. Enzymatic reactions have special requirements, some of which are surprising to the uninitiated, who may assume that biochemistry is always aqueous chemistry. Freemantle (2000) cites industrial research into the use of ionic liquids as media for enzyme-catalyzed reactions.

## 9.2   *ENVOI*

Much of what has been introduced in this book, and a great deal that is not, is described in Wypych's (2001) *Handbook of Solvents*. It contains information on uses of solvents in a wide variety of industries, toxicology and environmental effects of solvents, their safe use and disposal, recommendations for substitution of solvents by safer substances and processes, protection of workers from solvent exposure, detection and control of solvent residues in products, and regulations in effect in the United States, Canada, and the European Economic Community.

It concludes with a section on environmental contamination cleanup. Marcus's (1998) *The Properties of Solvents* also contains many tables of physical, chemical, optical, and electrochemical properties of solvents. The 260 solvents listed appear in each of the large tables in the same order, and are numbered so that it is easy to find all the available properties of a given substance. For example, values of 70 properties of DMSO are listed. Marcus also tabulates and discusses various applications, methods of purification, safety precautions and methods of disposal of solvents, and the significance of many of the parameters relevant to solvent effects, and their derivation.

These and other considerations are admirably discussed, and tables of relevant data provided, in the Appendix to Reichardt and Welton's (2011) book, *Solvents and Solvent Effects in Organic Chemistry*, to which reference has been repeatedly made in these chapters. This appendix also provides information on the toxicity of solvents and their safe handling, as well as methods of purification appropriate to various classes. We could not do better than to conclude by reiterating our recommendation of this book and especially its Appendix.

# APPENDIX

# (Tables listing parameters, selected values)

*Solvent Effects in Chemistry*, Second Edition. Erwin Buncel and Robert A. Stairs.
© 2016 John Wiley & Sons, Inc. Published 2016 by John Wiley & Sons, Inc.

**TABLE A.1  Properties of Selected Solvents (at 25°C Except as Noted)**

| Solvent | mp/°C | bp/°C | $\rho$/g cm⁻³ | $\eta$/cP | $\varepsilon_r$ | $n_D^{20}$ | Hazard (Fp)[a] | Disposal[b] |
|---|---|---|---|---|---|---|---|---|
| **Fluorocarbons** | | | | | | | | |
| f-n-Hexane[c] | −4 | 55–60 | 1.669 | | | 1.2515 | | A |
| f-Benzene | 4 | 82 | 1.612 | 1.200²⁰ | 2.01 | 1.3769 | (10) | D |
| **Hydrocarbons** | | | | | | | | |
| n-Hexane | −95 | 69 | 0.659 | 0.294 | 1.90 | 1.3749 | (−23) | D |
| Cyclohexane | 6.5 | 81 | 0.779 | 0.895 | 2.023²⁰ | 1.4260 | (−18) | D |
| Benzene | 5.5 | 80 | 0.874 | 0.601 | 2.27 | 1.5010 | CARC?, SK (−11) | D |
| Toluene | −93 | 111 | 0.867 | 0.550 | 2.38 | 1.4968 | (4) | D |
| **Halogenated hydrocarbons** | | | | | | | | |
| 1,1,2-Trichloro-f-ethane | −35 | 48 | 1.575 | | 2.48 | 1.3578 | | A |
| Dichloromethane | −97 | 40 | 1.325 | 0.425 | 8.9 | 1.4240 | POI, IRR | B |
| Trichloromethane | −63 | 61 | 1.492 | 0.542 | 4.7 | 1.4460 | CARC? | B |
| Tetrachloromethane | −23 | 77 | 1.594 | 0.880 | 2.23 | 1.4595 | POI, CARC? | B |
| 1,2-Dichloroethane | −35 | 83 | 1.256 | 0.39³⁰ | 10.36 | 1.4438 | CARC? (15) | D |
| Chlorobenzene | −45 | 132 | 1.107 | 0.80²⁰ | 5.62 | 1.5241 | POI | C |
| **Hydroxy-compounds** | | | | | | | | |
| Water | 0.0 | 100.0 | 0.9971 | 0.891 | 78.30 | 1.3330 | | |
| Water, crit. (supercrit. properties widely variable) | — | $t_c$ =374 (220 Bar) | 0.4 | | | 1.27(est) | (High pressure, temperature) | |
| Methanol | −98 | 64.8 | 0.7866 | 0.553 | 32.63 | 1.3284 | POI (11) | D |
| Ethanol | −114.5 | 78.3 | 0.7851 | 1.06 | 24.3 | 1.3610 | (8) | D |
| 1-Propanol | −126 | 97.2 | 0.7998 | 2.27 | 20.1 | 1.3854 | IRR (15) | D |
| 2-Propanol | −89 | 82.3 | 0.7854²⁰ | 1.77³⁰ | 18.3 | 1.3770 | (12) | D |
| 2-Methyl-2-propanol | 25.5 | 83 | 0.786 | 2.07 | 10.9³⁰ | 1.3860 | (4) | D |
| 1,2-Ethanediol | −13 | 198 | 1.113 | 16.2 | | 1.4310 | POI (110) | A |
| **Ethers** | | | | | | | | |
| Diethyl ether | −113 | 34.6 | 0.706 | 0.222 | 4.22 | 1.3506 | (−40) | D |
| Di-n-butyl ether | −98 | 143 | 0.764 | | | 1.3988 | (25) | D |
| Tetrahydrofuran | −108 | 67 | 0.886 | | 7.39 | 1.4070 | (−17) | D |

| Compound | | | | | | | | |
|---|---|---|---|---|---|---|---|---|
| 1,4-Dioxane | 11.8 | 102 | 1.034 | | 2.21 | 1.4215 | CARC? (12) | D |
| 1,2-Dimethoxyethane | −74 | 56 | 0.842 | | | 1.3923 | (0) | C |
| **Ketones** | | | | | | | | |
| Acetone (2-propanone) | −94 | 32 | 0.791 | 0.316 | 20.7 | 1.3585 | (−17) | D |
| 2-Butanone | −87 | 80 | 0.805 | | $18.5^{20}$ | 1.3788 | (−3) | D |
| **Esters** | | | | | | | | |
| Methyl formate | −99 | 32 | 0.974 | | $8.5^{20}$ | 1.3425 | (−27) | D |
| Ethyl acetate | −84 | 77 | 0.902 | 0.441 | 6.02 | 1.3720 | (−3) | D |
| γ-Butyrolactone | −45 | 205 | 1.120 | | | 1.4365 | (98) | C |
| Propylene carbonate | −55 | 240 | 1.189 | 2.53 | 65.1 | 1.4210 | (132) | A |
| **Nitrogenous compounds** | | | | | | | | |
| Ammonia | −77.7 | −33.4 | $0.65^{-10}$ | $0.25^{-33}$ | 16.9 | | COR | N |
| 2-Propylamine | −101 | 34 | 0.694 | 0.724 | $5.5^{20}$ | 1.3746 | COR, POI (−30 est) | D |
| Aniline | −6 | 184 | 1.022 | 3.71 | $6.89^{20}$ | 1.5863 | POI, SK, CARC? (70) | C |
| Acetonitrile | −48 | 82 | 0.786 | 0.345 | 36.2 | 1.3440 | LACH (5) | D |
| Benzonitrile | −13 | 188 | 1.010 | 1.24 | 26.0 | 1.5280 | IRR (71) | C |
| Pyridine | −42 | 115 | 0.978 | $0.954^{20}$ | 12.3 | 1.5102 | POI (20) | C |
| Formamide | 8.4 | 210 d | 1.129 | $3.76^{20}$ | 109.5 | | TER, IRR (??) | A |
| *N,N*-Dimethylformamide | −61 | 153 | 0.944 | 0.796 | 36.7 | 1.4305 | IRR (57) | C |
| Nitromethane | −29 | 101 | 1.127 | 0.608 | $38.6^{20}$ | 1.3820 | (35) | D |
| Nitrobenzene | 6 | 211 | 1.196 | $2.03^{20}$ | 34.6 | 1.5513 | POI, SK (87) | C |
| Hexamethylphosphoramide | 7 | 232 | 1.030 | $3.5^{60}$ | $30^{20}$ | 1.4579 | POI, CARC? (105) | A |
| Tetramethylurea | −1 | 177 | 0.971 | | | 1.4506 | (65) | C |
| **Sulfur compounds** | | | | | | | | |
| Dimethylsulfoxide | 18.4 | 189 | 1.101 | 1.98 | $47.6^{23}$ | 1.4787 | SK (95) | A |
| Sulfolane ((CH$_2$)$_5$SO$_2$) | 27 | 285 | 1.261 | $9.87^{30}$ | $44.0^{30}$ | 1.4840 | (165) | A |
| **Carboxylic acids** | | | | | | | | |
| Formic acid | 8.4 | 101 | 1.220 | $1.80^{20}$ | $57.9^{20}$ | 1.3704 | POI, COR (68) | C |
| Acetic acid | 16 | 117 | 1.059 | 1.16 | 6.19 | 1.3715 | COR, SK (40) | C |
| **Other acids** | | | | | | | | |
| Sulfuric acid | 10.5 | 330 d | 1.840 | 24.5 | 100 | | COR (−) | N |
| Phosphoric acid | 42 | 213 d | $1.834^{18}$ | 178 | ~61 | | COR (−) | N |

*(Continued)*

**TABLE A.1** (cont'd)

| Solvent | mp/°C | bp/°C | $\rho$/g cm$^{-3}$ | $\eta$/cP | $\varepsilon_r$ | $n_D^{20}$ | Hazard (Fp)[a] | Disposal[b] |
|---|---|---|---|---|---|---|---|---|
| Hydrogen fluoride | −83 | 19.4 | 0.987 | 0.24[6.25] | 60[19] | 1.2675[10] | POI, COR, SK (−) | F |
| Hydrogen cyanide | −14 | 26 | 0.699[20] | 0.20[20] | 106.8 | | POI (??) | |
| **Ionic liquids** | | | | | | | | |
| LiCl/NaCl 25:75 mol mol$^{-1}$ | 551 | | | | | | (−) | |
| LiCl/AlCl$_3$ 35:65 | 80 | | | | | | (−) | |
| [1-Ethyl-mim]$^+$ [TFSI]$^-$ [d] | −15 | | 1.52 | 34[20] | (E$_T$(30) = 52.6) | | (−) | |
| [1-Ethyl-mim]$^+$ [Al$_2$Cl$_7$]$^-$ | ca. −20 | | | 9.5[20] | | | (−) | |
| [1-Butyl-mim]$^+$[BF$_4$]$^-$ | −82 | | 1.15 | 219[20] | | | | |
| **Other inorganics** | | | | | | | | |
| Arsenic trifluoride | −8.5 | 63 | 2.67 | | 5.7 | | POI (−) | L |
| Thionyl chloride | −105 | 79 | 1.631 | | | 1.5140 | LACH (−) | N |
| Phosphorus oxychloride | 1.25 | 105.8 | 1.645 | | 13.9[22] | 1.460[25] | CORR (−) | N |
| Sulfur dioxide | −75.5 | −10 | 1.434[−10] | 0.428[−10] | 12.3[22] | 1.410 | POI, IRR (−) | K |
| Carbon disulfide | −112 | 46 | 1.266 | 0.363 | 2.64[20] | 1.6270 | POI (−33) | D |
| Carbon dioxide (crit.) | | $t_c$ = 31.1 (74 Bar) | 0.460 | | | | (High pressure) (−) | (vent) |

[a] POI, poisonous by inhalation or ingestion; IRR, irritating, especially to mucous membranes and eyes; SK, readily absorbed through the skin; LACH, lachrymator; CARC(?), carcinogenic (suspected); TER, teratogenic; CORR, Corrosive; (Fp), flash point, °C(where applicable).

[b] Waste-disposal methods adapted from those recommended in the annual catalogue of Aldrich Chemical Co. Ltd., 940 W. Saint Paul Ave., Milwaukee, WI 53233, USA:

(A) Mix with combustible diluent, incinerate. "Incinerate" here means burn in a chemical incinerator equipped with afterburner and scrubber.

(B) Ignite in the presence of sodium carbonate and slaked lime.

(C) Incinerate.

(D) Incinerate, with precautions due to its high inflammability.

(E) Add to excess of water, neutralize with sodium carbonate. Add sufficient calcium chloride to precipitate fluoride and carbonate. Dispose of solids to secure landfill.

(F) Consult supplier.

(G) Add to large excess water, precipitate as sulfide, neutralize, oxidize excess sulfide with hypochlorite. Solids to secure landfill, liquid to drain.

(H) Add to large excess of water, neutralize with slaked lime. Separate solids for disposal to secure landfill. Flush aqueous solution to drain.

[c] Perfluorinated substance designated by $f$.

[d] "mim" is 3-methylimidazolium; TFSI is bis(trifluoromethanesulfonyl)imide.

**TABLE A.2a    Solvent Property Parameters: Symmetric Properties**

Part 1: Solvatochromic parameters

| Symbol | Measure of (basis) | Reference |
|---|---|---|
| $a(^{14}N)$ | Polarity (ESR hyperfine splitting) | Knauer and Napier (1974) |
| $E_K$ | Polarity (energy of $d \to \pi^*$ transition in a Mo complex) | Walther (1974) |
| $E^*_{MLCT}$ | Similar, based on a W complex | Manuta and Lees (1986) |
| $E_T(30)$ | Polarity (CT abs freq of a dye) | Dimroth and Reichardt (1971) |
| $E_T^N$ | Normalized $E_T(30)$ | Reichardt and Harbusch-Görnert (1983) |
| $E_T^{SO}$ | Polarity ($n \to \pi^*$ abs freq of a sulfoxide) | Walter and Bauer (1977) |
| $G$ | Polarity (IR shifts) | Allerhand and Schleyer (1963) |
| $P$ | Polarity ($^{19}F$ nmr shifts) | Brownlee et al. (1972) |
| $P_y$ | Polarity ($\pi^* \to \pi$ emission of pyrene: relative intensities of two bands) | Dong and Winnik (1984) |
| $S$ | Polarity (composite) | Brownstein (1960) |
| $S'$ | Polarity (composite) | Drago (1992) |
| SPP | Polarity/polarizability (difference between UV-vis spectra of 2-(dimethylamino)-7-nitrofluorene and 2-fluoro-7-nitrofluorene) | Catalán et al. (1995) |
| $Z$ | Polarity (CT trans. of a pyridinium iodide) | Kosower (1958) |
| $\delta$ | Polarizability correction to $\pi^*$ | Kamlet et al. (1983) |
| $\theta_{1k}, \theta_{2k}$ | Polarity and polarizability components of $\pi^*$ | Sjöström and Wold (1981) |
| $\pi^*$ | Polarity/polarizability | Kamlet et al. (1977) |
| $\pi^*_{azo}$ | Polarity ($n \to \pi^*$ and $\pi \to \pi$ shifts in six azo-merocyanines) | Buncel and Rajagopal (1989, 1990) |
| $\chi_R, \chi_B$ | Polarity ($\pi \to \pi^*$ trans. of merocyanine dyes) | Brooker et al. (1965) |
| $\Phi$ | Polarity ($n \to \pi^*$ trans. in ketones) | Dubois and Bienvenüe (1968) |

Part 2: Parameters based on equilibrium (physical or chemical)

| Symbol | Measure of (basis) | References |
|---|---|---|
| $I$ | Gas-chromatographic retention index | Kováts (1961, 1965) |
| $P$ or $K_{o/w}$ (as $\log_{10}$) | Lyophilicity (partition between 1-octanol and water) | Hansch (1969) and Leo (1983) |
| KB | Kauri-butanol number (turbidity in solution of kauri resin in 1-butanol on addition of solvent) | ASTM (1982) |
| $L$ | Desmotropic constant (enol/diketo ratio of ethyl acetoacetate in the solvent) | Meyer (Meyer and Hopff, 1921) |

*(Continued)*

**TABLE A.2a**   (cont'd)

| $M$ | Miscibility number | Godfrey (1972) |
|---|---|---|
| $S$ (script) | Ionizing power ($\log k_2$ for a Menschutkin rn.) | Drougard and Decroocq (1969) |
| $X$ | Polarity (an $S_E2$ reaction) | Gielen and Nasielski (1967) |
| $Y$ | Ionizing power ($t$-butyl chloride solvolysis) | Grunwald and Winstein (1948) |
| $Y_X$ | Ionizing power (various solvolyses) | Various authors |
| $D_1$ | Conformational equilibrium shift | Eliel and Hofer (1973) |
| $\Omega$ | Polarity ([endo]/[exo] ratio in a Diels-Alder rn.) | Berson et al. (1962) |

| Part 3: Parameters based on other properties | | |
|---|---|---|
| Symbol | Measure of (basis) | References |
| $c$ | Cohesion pressure or cohesive energy density ($\Delta E_{vap}/V_m$) | Hildebrand and Scott (1950) |
| $g(\varepsilon_r)$ | Polarity/polarizability ( $= (\varepsilon_r - 1)/(2\varepsilon_r + 1)$; $\varepsilon_r$ is the relative permittivity (dielectric constant) | Kirkwood (1934) |
| $P$ | Polarizability $= (n^2 - 1)/(n^2 + 2)$ | Koppel and Pal'm (1972) (Lorentz-Lorenz, 1880) |
| $Q$ | Polarity: $R-P$ | Koppel and Pal'm (1972) |
| $R$ | Equal to $Y$: see below | Koppel and Pal'm (1972) |
| Sp | Solvophobic power (Gibbs energy of transfer of a nonpolar solute from water to the solvent) | Abraham et al. (1988) |
| $Y$ | Polarizability/polarity: $= (\varepsilon_r - 1)/(\varepsilon_r + 2)$ (cf. Kirkwood) | Koppel and Pal'm (1972) |
| | | Clausius (1879) and Mossotti (1850) |
| $\delta_H$ | Solubility parameter (square root of cohesive energy density $\Delta E_{vap}/V_m$) | Hildebrand and Scott (1950) |
| (No symbol) | Softness (used with Lewis acid/base strengths) | Pearson (1963) |
| ${}^1\chi^v/f$ | 1st-order valence molecular connectivity index | Kier (1981) |
| $S_{orb}$ | Hardness: $= 1/(E_{LUMO} - E_{HOMO})$ | Klopman (1968) |

**TABLE A.2b  Dual Parameters**

| Acid | Measure of (basis) | Basic | Measure of (basis) | References |
|---|---|---|---|---|
| $B$ | Nucleophilic solvation. (O–D IR stretch frequency difference, solvent—gas phase) | $E$ | Electrophilic solvation ($E_T(30)$ corrected for non-specific effects: polarity and polarizability) | Koppel and Pal'm (1972) and Koppel and Paju (1974) |
| $A_j$ | Acity: anion-solvating tendency (see $B_j$) | $B_j$ | Basity: cation-solvating tendency. (Bilinear correlation with $A_j$) of many effects.) | Swain et al. (1983) |
| $AN$ | Electron pair acceptor number ($^{31}$P NMR chemical shifts in triethylphosphine oxide) | $DN$ | Pair donor number ($-\Delta H$ of adduct formation with SbCl$_5$ in dilute solution in 1,2-dichloroethane) | Mayer et al. (1975, 1977) |
| $\alpha$ | HBD strength (enhanced solvatochromism of Reichardt's dye 30 relative to 4-nitroanisole) | $\beta$ | HBA strength (enhanced solvatochromism in HBA solvents for 4-nitroaniline relative to N,N-diethyl-4-nitroaniline) | Kamlet and Taft (1976) |
| $\alpha_1$ | HBD strength (part of $E_T(30)$ responding to solvent acidity) | $\beta_1$ | HBA strength ($^{19}$F NMR chemical shifts of 4-fluorophenol and 4-fluoroanisole) | Cerón-Carrasco et al. (2014a, b) and Laurence et al. (2014) |
| $C_A$ | Covalent part of acid strength (fit of enthalpy of adduct formation between EPD and EPA in solution to $-\Delta H = C_A C_B + E_A E_B$) | $C_B$ | Covalent part of base strength (see $C_A$) | Drago (1980); Drago and Wayland (1965) |
| $E_A$ | Electrovalent part of acid strength (see $C_A$) | $E_B$ | Electrovalent part of base strength (see $C_A$) | Drago (1980); Drago and Wayland (1965) |
|  |  | $N$ | Nucleophilicity (bilinear correlation (with $Y$, vide supra) of rates of solvolysis of t-butyl chloride) | Winstein et al. (1951, 1957) |
|  |  | $D_H$ | Hard basicity (Gibbs energy of transfer of Na$^+$, water to solvent) | Persson et al. (1987) |
|  |  | $D_S$ | Soft basicity (wavenumber shift of sym. stretch of HgBr$_2$) | Persson et al. (1987) |

(Continued)

**TABLE A.2b** (cont'd)

| Acid | Measure of (basis) | Basic | Measure of (basis) | References |
|---|---|---|---|---|
| $E_B^N$ | Lewis acidity (solvatochromism of $n \to \pi^*$ transition in 2,2,6,6,tetramethylpiperidine-1-oxide radical) | $D\pi$ | Soft basicity (retardation of [2+2] cycloaddition of diazodiphenylmethane to tetracyanoethene) | Oshima and Nagai (1985) |
| | | $\Delta H_{D-BF3}^0$ | Hard basicity (enthalpy of adduct formation between EPD solvent and $BF_3$ in dilute solution in dichloromethane) | Janowski et al. (1985); Mukerjee et al. (1982)<br>Maria and Gal (1985) |
| | | $M$ | Base softness (difference in Gibbs energies of transfer of $Ag^+$ and alkali metal ion from water to solvent) | Marcus (1987) |
| $A$ | Soft acidity (partial correlation of $\log K$ vs. $E_n$ for Lewis acid/base reactions of cations with ligand $Y^-$.) | $E_n$ | Soft basicity ($= E^o + 2.60$, where $E^o$ is the standard oxidation potential for $Y^- = \frac{1}{2}Y_2 + e^-$) | Edwards (1956) |
| $B$ | Hard (protonic) acidity (partial correlation of $\log K$, (as preceding) vs. $H$ | $H$ | $1.74 + pK$ for $HY = H^+ + Y^-$ | Edwards (1956) |
| $SA$ | Comparison of solvatochromism of an unhindered with a sterically hindered stilbazolium betaine dye. | $SB$ | Comparison of solvatochromism of 5-nitroindoline with N-methyl-5-nitroindoline | Catalán (2001) |
| $S_0^A$ | Softness of neutral acid (from Drago-Wayland parameters by rotation of coordinates) | $S_0^B$ | Softness of neutral base (from Drago-Wayland parameters by rotation) | See Section 3.9 |
| $S_+^A$ | Softness of cations as acids (from Edwards's parameters, by rotation) | | | See Section 3.9 |

TABLE A.3  Values of Selected Parameters for Selected Solvents (in order of increasing $E_T^N$)

|  | Solvents | 1 $R_v$ | 2 $Q_v$ | 3 $\beta_\mu^{1/2}$ | 4 $\delta_H$ | 5 $E_T^N$ | 6 $Z$ | 7 $A_j$ | 8 $B_j$ | 9 $\alpha$ | 10 $\beta$ | 11 AN | 12 DN |
|---|---|---|---|---|---|---|---|---|---|---|---|---|---|
| 1 | Cyclohexane | 0.256 | 0.00 | 0.00 | 16.9 | 0.006 | 60.1 | 0.02 | 0.06 | 0.00 | 0.00 | — | 4.8 |
| 2 | n-Hexane | 0.229 | 0.00 | 0.00 | 14.9 | 0.009 | 50.0 | 0.01 | −0.01 | 0.00 | 0.00 | 0.0 | 5.3 |
| 3 | Triethylamine | 0.243 | 0.09 | 1.28 | 15.2 | 0.043 | 50.0 | 0.08 | 0.19 | 0.00 | 0.71 | 1.43 | 0.5 |
| 4 | Tetrachloromethane | 0.274 | 0.03 | 0.00 | 17.6 | 0.052 | 52.0 | 0.09 | 0.34 | 0.00 | 0.10 | 8.6 | 9.6 |
| 5 | Carbon disulfide | 0.355 | 0.00 | 0.00 | 20.4 | 0.065 | 52.0 | 0.10 | 0.38 | 0.00 | 0.07 | — | — |
| 6 | Toluene | 0.293 | 0.03 | 0.50 | 18.2 | 0.099 | 54.0 | 0.13 | 0.54 | 0.00 | 0.11 | — | — |
| 7 | Benzene | 0.294 | 0.00 | 0.00 | 18.8 | 0.111 | 54.0 | 0.15 | 0.59 | 0.00 | 0.10 | 8.2 | — |
| 8 | Diethyl ether | 0.216 | 0.32 | 1.95 | 15.4 | 0.117 | 55.0 | 0.12 | 0.34 | 0.00 | 0.47 | 3.9 | 19.2 |
| 9 | 1,4-Dioxane | 0.254 | 0.04 | 0.85 | 20.5 | 0.164 | 64.6 | 0.19 | 0.67 | 0.00 | 0.37 | 10.8 | 14.8 |
| 10 | Chlorobenzene | 0.306 | 0.31 | 2.80 | 19.4 | 0.188 | 58.0 | 0.20 | 0.65 | 0.00 | 0.07 | — | — |
| 11 | Tetrahydrofuran | 0.246 | 0.44 | 3.37 | 19.0 | 0.207 | 58.8 | 0.17 | 0.67 | 0.00 | 0.55 | 8.0 | 20.0 |
| 12 | Ethyl acetate | 0.227 | 0.40 | 3.22 | 18.4 | 0.228 | 59.4 | 0.21 | 0.59 | 0.00 | 0.45 | 9.3 | 17.1 |
| 13 | Trichloromethane | 0.267 | 0.30 | 2.21 | 18.9 | 0.259 | 63.2 | 0.42 | 0.73 | 0.20 | 0.10 | 23.1 | — |
| 14 | Ammonia | 0.201 | 0.640 | 5.55 | — | (0.26) | — | — | — | — | — | — | — |
| 15 | 4-Methylpyridine | 0.296 | — | 1.3 | 20.8 | 0.272 | — | — | — | 0.00 | 0.67 | — | — |
| 16 | Pyridine | 0.299 | 0.50 | 4.60 | 21.6 | 0.302 | 64.0 | 0.24 | 0.96 | 0.00 | 0.64 | 14.2 | 33.1 |
| 17 | Dichloromethane | 0.255 | 0.47 | 3.40 | 20.3 | 0.309 | 64.2 | 0.33 | 0.80 | 0.13 | 0.10 | 20.4 | — |
| 18 | Hexamethylphosphoramide | 0.273 | 0.63 | 7.31 | 18.3 | 0.315 | 62.8 | 0.00 | 1.07 | 0.00 | 1.05 | 9.8 | 38.8 |
| 19 | 1,2-Dichloroethane | 0.266 | 0.50 | 3.56 | 20.3 | 0.327 | 63.4 | 0.30 | 0.82 | 0.00 | 0.10 | 16.7 | 0.0 |
| 20 | 2-Butanone | 0.231 | 0.63 | 5.08 | 19.0 | 0.327 | 64.0 | 0.23 | 0.74 | 0.00 | 0.48 | — | 17.4 |
| 21 | Sulfur dioxide | 0.248 | 0.542 | 1.48 | 22.8 | (0.33) | 65.0 | — | — | — | — | — | — |
| 22 | Benzonitrile | 0.308 | 0.585 | 7.19 | 22.8 | 0.333 | 65.5 | 0.25 | 0.81 | 0.00 | 0.37 | 15.5 | 11.9 |
| 23 | Acetone | 0.221 | 0.65 | 5.49 | 19.7 | 0.355 | 65.5 | 0.30 | 0.93 | 0.08 | 0.43 | 12.5 | 17.0 |
| 24 | N,N-Dimethylformamide | 0.258 | 0.66 | 6.43 | 24.0 | 0.386 | 68.4 | 0.30 | 0.93 | 0.00 | 0.69 | 16.0 | 26.6 |
| 25 | t-Butanol | 0.235 | 0.533 | 2.96 | — | 0.389 | 71.3 | 0.45 | 0.50 | 0.42 | 0.93 | 27.1 | — |

(Continued)

**TABLE A.3**  (cont'd)

| | Solvents | 1 $R_v$ | 2 $Q_v$ | 3 $\beta_\mu^{1/2}$ | 4 $\delta_H$ | 5 $E_T^N$ | 6 $Z$ | 7 $A_j$ | 8 $B_j$ | 9 $\alpha$ | 10 $\beta$ | 11 AN | 12 DN |
|---|---|---|---|---|---|---|---|---|---|---|---|---|---|
| 26 | Dimethylsulfoxide | 0.283 | 0.66 | 8.38 | 26.6 | 0.444 | 70.2 | 0.34 | 1.08 | 0.00 | 0.76 | 19.3 | 29.8 |
| 27 | Acetonitrile | 0.212 | 0.71 | 8.53 | 24.2 | 0.460 | 71.3 | 0.37 | 0.86 | 0.19 | 0.40 | 18.9 | 14.1 |
| 28 | Nitromethane | 0.232 | 0.68 | 8.49 | 25.8 | 0.481 | 71.2 | 0.39 | 0.92 | 0.22 | 0.06 | 20.5 | 2.7 |
| 29 | 2-Propanol | 0.230 | 0.63 | 3.29 | — | 0.546 | 76.3 | 0.59 | 0.44 | 0.76 | 0.84 | 33.5 | — |
| 30 | 1-Propanol | 0.235 | 0.63 | 3.32 | — | 0.617 | 78.3 | 0.63 | 0.44 | 0.86 | 0.90 | 33.7 | — |
| 31 | Hydrogen cyanide | 0.168 | 0.804 | 2.90 | — | (0.65) | — | — | — | — | — | — | — |
| 32 | Ethanol | 0.221 | 0.66 | 3.97 | — | 0.654 | 79.6 | 0.66 | 0.45 | 0.86 | 0.75 | 37.1 | — |
| 33 | Methanol | 0.204 | 0.71 | 4.68 | — | 0.762 | 83.6 | 0.75 | 0.50 | 0.98 | 0.66 | 41.5 | 18.4 |
| 34 | 1 2-Ethanediol | 0.257 | 0.67 | 4.14 | — | 0.790 | 85.1 | 0.78 | 0.84 | 0.90 | 0.52 | — | — |
| 35 | Hydrogen fluoride | 0.102 | 0.863 | 7.39 | — | (0.83) | — | — | — | — | — | — | — |
| 36 | 2.2.2-Trifluoroethanol | 0.182 | 0.712 | | — | 0.898 | — | — | — | 1.51 | 0.00 | 53.8 | — |
| 37 | Water | 0.206 | 0.76 | 7.27 | — | 1.000 | 94.6 | 1.00 | 1.00 | 1.17 | 0.47 | 54.8 | — |

| | 13 | 14 | 15 | 16 | 17 | 18 | 19 | 20 | 21 | 22 | 23 | 24 | 25 | 26 |
|---|---|---|---|---|---|---|---|---|---|---|---|---|---|---|
| | $E_b$ | $C_b$ | $\pi^*$ | $\pi^*_{azo}$ | $a(^{14}N)$ | $S$ | $\chi_R$ | $S$ | $W$ | $\Omega$ | $\mu$ | $\log K_{o/w}$ | $S_0^a$ | $S_0^a$ |
| 1 | — | — | 0.00 | 0.00 | — | -0.324 | 50.0 | -4.15 | — | 0.595 | — | 3.44 | — | — |
| 2 | — | — | -0.11 | -0.09 | 1.513 | -0.337 | 50.9 | — | — | — | — | 4.11 | — | — |
| 3 | 1.32 | 5.73 | 0.09 | 0.05 | — | -0.285 | 49.3 | -2.85 | — | 0.445 | 0.150 | 1.64 | 0.65 | — |
| 4 | — | — | 0.21 | 0.15 | 1.533 | -0.245 | 48.7 | — | — | — | — | 2.64 | — | — |
| 5 | — | — | 0.51 | 0.25 | 1.529 | -0.240 | — | — | — | — | — | — | — | — |
| 6 | — | — | 0.49 | 0.38 | 1.535 | -0.237 | 47.2 | — | — | — | — | 2.69 | — | — |
| 7 | 0.70 | 0.45 | 0.55 | 0.40 | 1.540 | -0.215 | 46.9 | -1.74 | — | 0.497 | 0.350 | 2.13 | 0.70 | — |
| 8 | 1.80 | 1.63 | 0.24 | 0.16 | 1.533 | -0.277 | 48.3 | -2.92 | — | 0.466 | 0.050 | 0.89 | -0.27 | — |
| 9 | 1.86 | 1.29 | 0.49 | 0.34 | 1.545 | -0.179 | 48.4 | -1.43 | -7.3 | — | 0.050 | -0.27 | -0.37 | — |
| 10 | — | — | 0.68 | 0.58 | 1.547 | -0.182 | 45.2 | -1.15 | — | — | 0.250 | 2.80 | — | — |
| 11 | 1.64 | 2.18 | 0.55 | 0.40 | 1.537 | — | 46.6 | -1.54 | -6.073 | — | — | — | -0.05 | — |
| 12 | 1.62 | 0.98 | 0.45 | 0.37 | — | -0.210 | 47.2 | -1.66 | -5.947 | — | — | 0.73 | -0.16 | — |
| 13 | — | — | 0.69 | 0.62 | 1.586 | -0.200 | 44.2 | -0.89 | — | — | 0.150 | 1.97 | -1.13 | — |
| 14 | 1.78 | 3.54 | 0.87 | 0.80 | 1.561 | -0.197 | 43.9 | — | -4.670 | 0.595 | 0.640 | 0.65 | -0.04 | — |
| 15 | — | — | 0.73 | 0.62 | 1.575 | -0.189 | 44.9 | -0.55 | — | — | — | 1.25 | — | — |
| 16 | — | — | 0.87 | 0.85 | — | — | — | — | — | — | — | — | — | — |
| 17 | — | — | — | 0.63 | 1.566 | -0.151 | — | -0.42 | — | 0.600 | 0.290 | 1.48 | — | — |
| 18 | — | — | 0.67 | 0.61 | — | — | — | — | — | — | 0.030 | — | — | — |
| 19 | 1.74 | 1.26 | 0.62 | 0.53 | 1.553 | -0.175 | 45.7 | -0.82 | -5.067 | 0.619 | 0.030 | -0.24 | -0.25 | — |
| 20 | 2.19 | 1.31 | 0.88 | 0.86 | 1.564 | -0.142 | 43.7 | -0.22 | -4.298 | 0.620 | 0.110 | — | -0.70 | — |
| 21 | 2.40 | 1.47 | 1.00 | 1.00 | 1.569 | — | 42.0 | — | -3.738 | — | 0.220 | -1.35 | -0.89 | — |
| 22 | 1.64 | 0.71 | 0.66 | 0.63 | 1.567 | -0.104 | 45.7 | -0.33 | -4.221 | 0.692 | 0.350 | -0.34 | -0.21 | — |
| 23 | — | — | 0.75 | 0.70 | 1.576 | -0.134 | 44.0 | 0.04 | -3.921 | 0.680 | 0.030 | — | — | — |
| 24 | — | — | 0.48 | 0.51 | 1.597 | -0.041 | 44.5 | — | -3.970 | — | — | 0.05 | — | -0.52 |
| 25 | — | — | 0.52 | 0.52 | — | -0.016 | 44.1 | — | — | — | 0.160 | 0.25 | — | — |

*(Continued)*

**TABLE A.3** (cont'd)

| | 13 | 14 | 15 | 16 | 17 | 18 | 19 | 20 | 21 | 22 | 23 | 24 | 25 | 26 |
|---|---|---|---|---|---|---|---|---|---|---|---|---|---|---|
| | $E_b$ | $C_b$ | $\pi^*$ | $\pi^*_{azo}$ | $a(^{14}N)$ | $S$ | $X_R$ | $S$ | $W$ | $\Omega$ | $\mu$ | $\log K_{o/w}$ | $S_0^a$ | $S_0^a$ |
| 26 | 1.85 | 1.09 | 0.54 | 0.54 | 1.603 | 0.000 | 43.9 | -2.02 | -3.204 | 0.718 | 0.080 | -0.31 | -0.38 | -0.84 |
| 27 | 1.80 | 0.65 | 0.60 | 0.60 | 1.621 | 0.050 | 43.1 | -1.89 | -2.796 | 0.845 | 0.020 | -0.77 | -0.38 | -0.72 |
| 28 | — | — | 0.92 | 0.62 | 1.636 | 0.068 | 40.4 | — | — | — | -0.030 | -1.36 | — | — |
| 29 | — | — | 1.09 | 1.03 | 1.717 | 0.154 | — | — | -1.180 | 0.869 | 0.000 | — | — | -1.37 |
| 30 | — | — | — | — | — | — | — | — | — | — | — | — | — | -1.76 |
| 31 | — | — | 0.73 | — | — | — | — | — | — | — | — | — | — | -1.13 |
| 32 | — | — | 0.41 | — | 1.586 | — | — | — | -0.105 | — | — | — | — | -0.97 |
| 33 | — | — | — | — | — | — | — | — | — | — | — | — | — | 0.03 |
| 34 | 2.31 | 2.04 | — | — | — | — | — | — | — | — | — | — | -0.74 | — |
| 35 | 1.75 | 0.62 | — | — | — | — | — | — | — | — | — | — | -0.33 | — |
| 36 | 1.74 | 3.93 | — | — | — | — | — | — | — | — | — | — | 0.04 | — |
| 37 | 1.19 | 0.10 | — | — | — | — | — | — | — | — | — | — | 0.17 | — |

Values in parenthesis of $E_T^N$ for HF, SO$_2$, and NH$_3$ were assigned by guided guesswork, and are not to be trusted. That for HCN was estimated by extrapolation from linear aliphatic nitriles (see Section 2.10).

# ANSWERS

NOTE: Answers to selected problems

## CHAPTER 1

**1.1** (c) $\Delta G^0_{373} = -3.09\,\text{kJ}\,\text{mol}^{-1}$; $K_{373} = 2.7\,\text{l}^2\,\text{mol}^{-1}$; $f = 0.62$.

**1.2** (a) $K = 89.6$ (No, no units). (b) $K_{x(\text{ideal})} = K\ 36.6$.

**1.3** $f = 1.24$.

**1.4** $E_a = 42.1\,\text{kJ}\,\text{mol}^{-1}$; $PZ = 8 \times 10^{10}$; $\Delta_{\ddagger}H^0_{298} = 39.6\,\text{kJ}\,\text{mol}^{-1}$; $\Delta_{\ddagger}G^0_{298} = 69.9\,\text{kJ}\,\text{mol}^{-1}$;
$\Delta_{\ddagger}S^0_{298} = -102\,\text{J}\,\text{K}^{-1}\,\text{mol}^{-1}$.

$Z = 2.8 \times 10^8$. $P \sim 0.3$.

## CHAPTER 2

**2.1** $K_x = 46.1$.

**2.2** $C < 10^{-4}$ M, approximately.

**2.3** (iii) $KC$ large, order $\rightarrow 1$; $KC$ small, order $\rightarrow 2$.
(iv) Order $= 3/2$ when $KC = 0.75$.

**2.4** $K_1 = 24.7$; $K_2 = 6.2$ (dimensionless).

*Solvent Effects in Chemistry*, Second Edition. Erwin Buncel and Robert A. Stairs.
© 2016 John Wiley & Sons, Inc. Published 2016 by John Wiley & Sons, Inc.

## CHAPTER 3

**3.2**  $pK_a = -1.74$.

**3.3**  $K_a = (4.05 \pm 0.04) \times 10^{-6}$.

## CHAPTER 6

**6.1**  $\Delta_{tr}G = 15.1 \ \text{kJ} \, \text{mol}^{-1}$

**6.1**  $\Delta_{tr}G\left(A^{\ddagger}\right) = 6.7 \, \text{kJ} \, \text{mol}^{-1}$

# REFERENCES

Abbott AP, Capper G, Davies DL, Rasheed RK, Tambyraja Y. *Chem Commun* 2003:70–71.

Abbott AP, Capper G, Davies DL, Rasheed RK. *Inorg Chem* 2004;**43**:3447–3452.

Abboud J-LM, Notario R. *Pure Appl Chem* 1999;**71** (4):645–718.

Abboud J-LM, Taft RW. *J Phys Chem* 1979;**83**:412–419.

Abboud J-LM, Notario R, Bertrán J, Solà M. *Prog Phys Org Chem* 1993;**19**:1–182.

Abraham MH. *Prog Phys Org Chem* 1974;**11**:1–87.

Abraham MH. *J Am Chem Soc* 1982;**104**:2085–2094.

Abraham MH, Grellier PL, McGill RA. *J Chem Soc, Perkin Trans II* 1988:339–345.

Adams C, Earle M, Roberts G, Seddon K. *Chem Commun* 1998; (19):2097–2098.

Ahrland S, Chatt J, Davies NR. *Q Rev* 1958;**12**:265–276.

Alfonsi K, Colberg J, Dunn PJ, Fevig T, Jennings S, Johnson TA, Kleine HP, Knight C, Nagy MA, Perry DA, Stefaniak M. *Green Chem* 2008;**10**:31–36.

Allerhand A, Schleyer P v R. *J Am Chem Soc* 1963;**85**:371–380.

Alvarez FJ, Schowen RL. In: Buncel E, Lee CC, editors. *Isotopes in Organic Chemistry*. Volume **7**, Amsterdam: Elsevier; 1987. p 1–60.

American Society for Testing and Materials. *Annual Book of ASTM Standards 1982*. Philadelphia (PA): The Society; 1982. p 142–144.

Amis ES, Hinton JF. *Solvent Effects on Chemical Phenomena*. Volume **1**, New York/London: Academic Press; 1973.

Amis ES, LaMer VK. *J Am Chem Soc* 1939;**61**:905–913.

Amis ES, Price JE. *J Phys Chem* 1943;**47**:338–348.

Anonymous. Fluorous Technologies Limited. Pittsburgh (PA). Available at http//www. fluorous.com/flle.php. Accessed February 11, 2015.

Anonymous. Available at http//www.sust-chem.ethz.ch/tools/ecosolvent. Accessed February 11, 2015.

Arnot JA, Mackay D. *Environ Sci Technol* 2008;**42** (13):4648–4654.

Arshadi M, Yamdagni R, Kebarle P. *J Phys Chem* 1970;**74**:1475–1482.

Ashbaugh HS, Paulaitis ME. *J Phys Chem* 1996;**100**:1900–1913.

Atkins PW. *Physical Chemistry*. 6th ed. Oxford/New York: Oxford University Press/Freeman; 1998.

Atkins PW, de Paula J. *Physical Chemistry*. 9th ed. Oxford/New York: Oxford University Press/Freeman; 2010.

Audrieth LF, Kleinberg J. *Non-Aqueous Solvents*. New York/London: John Wiley & Sons, Inc./Chapman and Hall; 1953.

Bagno A. *Winter School on Organic Chemistry*. Bressanone:; 1998.

Balakrishnan VK, Dust JM, vanLoon GW, Buncel E. *Can J Chem* 2001;**79**:157–173.

Barrow GM. *Physical Chemistry*. 5th ed. New York: McGraw-Hill; 1988. p 327–340.

Barthel J, Gores H-J. In: Mamantov G, Popov AI, editors. *Chemistry of Nonaqueous Solutions: Current Progress*. New York/Weinheim/Cambridge: VCH; 1994. p 6 11.

Bartlett PDJ. *Am Chem Soc* 1972;**94**:2161–2170.

Barton AFM. *Chem Rev* 1975;**75**:731–753.

Basolo F, Pearson RG. *Mechanisms of Inorganic Reactions*. 2nd ed. New York: John Wiley & Sons, Inc.; 1967.

Bassett J, Denney RC, Jeffrey GH, Mendham J. *Textbook of Quantitative Inorganic Analysis*. Harlow: Longman; 1978. p 687–688.

Bates ED, Mayton RD, Ntai I, Davis JH Jr. *J Am Chem Soc* 2002;**124**:926–927.

Baudequin C, Brégeon D, Levillain J, Guillen F, Plaquevent J-C, Gaumont A-C. *Tetrahedron* 2005;**16**:3921–3945.

Bell RP. *Acid-base catalysis*. Oxford: Oxford University Press; 1941.

Bell RP. *The Proton in Chemistry*. 2nd ed. Ithaca (NY): Cornell University Press; 1973.

Bell RP, Higginson WCE. *Proc Roy Soc Lond A* 1949;**197**:141–159.

Bell RP, Bascombe KN, McCoubrey JC. *J Chem Soc* 1956:1286–1291.

Bentley TW, Roberts K. *J Org Chem* 1985;**50**:4821–4828.

Berry RS, Rice SA, Ross J. *Physical Chemistry*. New York/Oxford: Oxford University Press; 2000.

Berson JA, Hamlet Z, Mueller WA. *J Am Chem Soc* 1962;**84**:297–304.

Berthelot PEM, Pean de St. Gilles L. *Ann Chim Phys* 1862;**65** (3):385 **66**, 5.

Bika K, Gaertner P. Applications of chiral ionic liquids. *Eur J Org Chem* 2008:3235.

Billmeyer FW Jr. *Textbook of Polymer Science*. 2nd ed. New York: Wiley-Interscience; 1971. p 289 358–364.

Bjerrum N. *Kgl Danske Vidensk Selskab* 1926;**7** (9 See also Davies (1962), or Harned and Owen (1950)).

Blagoeva IB, Toteva MM, Ouarti N, Ruasse M-P. *J Org Chem* 2001;**66** (6):2123–2130.

Blandamer MJ. In: Gold V, Bethell D, editors. Volume **14**, London: Academic Press; 1977. *Advances in Physical Organic Chemistry*; p 203–343.

Blandamer MJ, Burgess J. *Pure Appl Chem* 1982;**54** (12):2285–2296.

Bloom H, Hastie JW. In: Waddington TC, editor. *Non-Aqueous Solvent Systems*. London/New York: Academic Press; 1965. p 353ff.

Bochevarov AD, Harder E, Hughes TF, Greenwood JR, Braden DA, Philipp DM, Rinaldo D, Halls MD, Zhang J, Friesner RA. *Int J Quantum Chem* 2013;**113** (18):2110–2142.

Bohme DK, Mackay GI. *J Am Chem Soc* 1981;**103**:978–979.

Boon J, Levitsky J, Pflug J, Wilkes J. *J Org Chem* 1986;**51**:480–483.

Bordwell FG. *Acc Chem Res* 1988;**21**:456–463.

Born M. *Z Phys* 1920;**1**:45–48.

Born M, Green MS. *A General Theory of Liquids*. Cambridge: Cambridge University Press; 1949.

Bosnich B. *J Am Chem Soc* 1967;**89**:6143–6148.

Bowden K, Cook RS. *J Chem Soc B* 1968:1529–1533.

Bowman NS, McCloud GT, Schweitzer GK. *J Am Chem Soc* 1968;**90**:3848–3852.

Brønsted JN, Pederson KJ. *Z Phys Chem* 1924;**108**:185.

Brooker LGS, Craig AC, Heseltine DW, Jenkins PW, Lincoln LL. *J Am Chem Soc* 1965;**87**:2443–2450.

Brownlee RTC, Dayel SK, Lyle JL, Taft RW. *J Am Chem Soc* 1972;**94**:7208–7209.

Brownstein S. *Can J Chem* 1960;**38**:1590–1596.

Buckingham RA. *Proc Roy Soc Lond A* 1938;**168**:264–283.

Buncel E. *Acc Chem Res* 1975a;**8**:132–139.

Buncel E. *Carbanions. Mechanistic and Isotopic Aspects*. Amsterdam: Elsevier; 1975b. p 16.

Buncel E. *Can J Chem* 2000;**78**:1251–1271.

Buncel E, Dust JM. *Carbanion Chemistry. Structures and Mechanisms*. Washington (DC): American Chemical Society; 2003.

Buncel E, Lawton BT. *Can J Chem* 1965;**43**:862–875.

Buncel E, Menon BC. *J Org Chem* 1979;**44**:317–320.

Buncel E, Millington JP. *Can J Chem* 1965;**43**:556–564.

Buncel E, Rajagopal S. *J Org Chem* 1989;**54**:798–809.

Buncel E, Rajagopal S. *Acc Chem Res* 1990;**23**:226–231.

Buncel E, Rajagopal S. *Dyes Pigm* 1991;**17**:303.

Buncel E, Saunders WH, editors. *Isotopes in Organic Chemistry*. Volume **8**, Amsterdam: Elsevier; 1992.

Buncel E, Symons EA. *J Am Chem Soc* 1976;**98**:656–660.

Buncel E, Symons EA. In: Bertini I, Lunazzi L, Dei A, editors. *Advances in Solution Chemistry*. New York: Plenum Press; 1981. p 355–371.

Buncel E, Symons EA. In: Zuckerman JJ, editor. *Inorganic Reactions and Methods*. Volume **1**, Weinheim: VCH; 1986. p 69–75.

Buncel E, Um IH. *J Chem Soc Chem Commun* 1986;**8**:595.

Buncel E, Um IH. *Tetrahedron* 2004;**60**:7801–7825.

Buncel E, Wilson H. *Adv Phys Org Chem* 1977;**14**:133–202.

Buncel E, Wilson H. *Acc Chem Res* 1979;**12**:42–48.

Buncel E, Wilson H. *J Chem Educ* 1980;**57**:629–633.

Buncel E, Symons EA, Zabel AW. *Chem Commun* 1965:173–174.

Buncel E, Menon BC, Colpa JP. *Can J Chem* 1979;**57**:999–1005.

Buncel E, Chuaqui C, Wilson H. *J Org Chem* 1980;**45**:3621–3626.

Buncel E, Chuaqui C, Wilson H. *Int J Chem Kinet* 1982;**14**:823–837.

Buncel E, Boone C, Lely HA. *Inorg Chim Acta* 1986;**125**:167–172.

Buncel E, Davey JP, Jones JR, Perring KD. *J Chem Soc Perkin Trans* 1990;**2**:169–173.

Buncel E, Dust JM, Terrier F. *Chem Rev* 1995;**95**:2261–2280.

Bunnett JF, Olsen FP. *Can J Chem* 1966;**44**:1899–1916. 1917–1931.

Burrell H. *Interchem Rev* 1955;**14** (3–16):31–46.

Caldin EF, Grant MW. *J Chem Soc Faraday Trans I* 1973;**69**:1648–1654.

Caldin EF, Hasinoff BB. *J Chem Soc Faraday Trans I* 1975;**71**:515–527.

Capello C, Hellweg S, Hungerbühler J. *The Ecosolvent Tool*. Zurich: ETH Zurich, Safety & Environmental Technology Group; 2006.

Capello C, Fischer U, Hungerbühler J. *Green Chem* 2007;**9**:927–934.

Car R, Parrinello M. *Phys Rev Lett* 1985;**55**:2471–2474.

Carey FA. *Organic Chemistry*. 3rd ed. New York: McGraw Hill; 1996.

Carlson R, Lundstedt T, Albano C. *Acta Chem Scand B* 1985;**39**:79–91.

Carmichael H. *Chem Br* 2000;January:36–38.

Carroll FA. *Perspectives on Structure and Mechanism in Organic Chemistry*. Pacific Grove (CA): Brooks/Cole Publishers; 1998.

Castro EA. *Chem Rev* 1999;**99**:3505–3524.

Catalán J. In: Wypych G, editor. *Handbook of Solvents*. Toronto/New York: ChemTec; 2001. p 583–616.

Catalán J, López V, Pérez P, Martin-Villamil R. *Liebigs Ann* 1995:241–252.

Cerfontain H. *Mechanistic Aspects in Aromatic Sulfonation and Desulfonation*. New York: John Wiley & Sons, Inc; 1968.

Cerón-Carrasco JP, Jacquemain D, Laurence C, Planchat A, Reichardt C, Sraïdi K. *J Phys Org Chem* 2014a;**27**:512–518.

Cerón-Carrasco JP, Jacquemain D, Laurence C, Planchat A, Reichardt C, Sraïdi K. *J Phys Chem B* 2014b;**118**:4605–4614.

Chandler D, Andersen HC. *J Chem Phys* 1972;**57**:1930–1937.

Chandrasekhar J, Jorgensen WL. *J Am Chem Soc* 1985;**107**:2974–2975.

Chapman NB, Shorter J, editors. *Advances in Linear Free Energy Relationships; Correlation Analysis in Chemistry-Recent Advances*. London: Plenum; 1972.

Chau PL, Forester TR, Smith W. *Mol Phys* 1996;**89**:1033–1055.

Chauvin Y, Mussman L, Olivier H. *Angew Chem Int Ed Engl* 1995;**34**:2698–2700.

Chen T, Hefter G, Marcus Y. *J Solution Chem* 2000;**29**:201–216.

Chum HL, Osteryoung RA. In: Inman D, Lovering DG, editors. *Ionic Liquids*. New York/London: Plenum; 1981. p 407–423.

Clark RJ, Brimm EO. *Inorg Chem* 1965;**4**:651–654.

Clayden J, Greeves N, Warren S, Wothers P. *Organic Chemistry*. Oxford: Oxford University Press; 2001.

Coetzee JF, Deshmukh BK. *Chem Rev* 1990;**90**:827–835.

Coetzee JF, Ritchie CD, editors. *Solute-Solvent Interactions*. New York/London: Marcel Dekker; 1969.

Collard DM, Jones AG, Kriegel RM. *J Chem Educ* 2001;**78**:70–72.

Collins CJ, Bowman NS, editors. *Isotope Effects in Chemical Reactions*. New York: Van Nostrand Reinhold; 1970.

Conway BE. *Electrochemical Data*. Amsterdam: Elsevier; 1952 reprinted 1969, Greenwood, Wesrport, CT.

Cook GK, Meyer JM. *J Am Chem Soc* 1995;**117**:7139–7156.

Covington AK, Newman KE. In: Furter WF, editor. *Thermodynamic Behaviour of Electrolytes in Mixed Solvents*. Volume **155**, Washington (DC): American Chemical Society; 1976. Advances in Chemistry Series; p 153–196.

Cox BG, Webster KC. *J Chem Soc* 1936:1635–1637.

Cox BG. *Ann Rep Chem Soc A* 1973;**70**:249–274.

Cox BG. *Modern Liquid Phase Kinetics*. Oxford: Oxford University Press; 1994.

Cox BG. *Acids and Bases: Solvent effects on Acid-Base Strength*. Oxford: Oxford University Press; 2013.

Cox RA, Yates K. *J Am Chem Soc* 1978;**100**:3861–3867.

Cox RA, Yates K. *Can J Chem* 1981;**59**:2116–2124.

Cox RA, Yates K. *Can J Chem* 1983;**61**:2225–2243.

Cox RA. *Acc Chem Res* 1987;**20**:27–31.

Cox RA. *Adv Phys Org Chem* 2000;**35**:1–66.

Coxeter HSM. *Introduction to Geometry*. New York/London: John Wiley & Sons, Inc. p. 128, Eqn. 8. 84; 1961.

Cram DJ. *Fundamentals of Carbanion Chemistry*. New York: Academic Press Chapters 2–4; 1965.

Cram DJ, Cram JM. *Acc Chem Res* 1978;**11**:8–14.

Cram DJ, Rickborn B, Knox GR. *J Am Chem Soc* 1960;**82**:6412–6413.

Cukierman S. *Biochim Biophys Acta* 2006;**1757**:876–878.

Davidson ER, Feller D. *Chem Rev* 1986;**86**:681–686.

Davies CW. *Ion Association*. London: Butterworths; 1962.

Davies CW, James JC. *Proc R Soc* 1948;**195A**:116–123.

Debye PJW, Hückel E. *Z Phys* 1923;**24**:185–206.

De Grotthuss CJT. *Ann Chim* 1806;**58**:54–73.

de Rege PJF, Gladyz JA, Horváth I. *Science* 1997;**276**:776–779.

Desrosiers N, Desnoyers JE. *Can J Chem* 1976;**54**:3800–3808.

Dewar MJS. In: Seeman JI, editor. *A Semiempirical Life*. Washington (DC): American Chemical Society; 1992.

Dewar MJS, Storch DM. *J Chem Soc Chem Commun* 1985:94–96.

Dewar MJS, Storch DM. *J Chem Soc Perkin Trans* 1989;**2**:877–885.

Dewar MJS, Thiel WJ. *J Am Chem Soc* 1977;**99**:4899–4906. 4907–4917.

Dimroth K, Reichardt C. *Justus Liebigs Ann Chem* 1969;**727**:93–105.

Dimroth K, Reichardt C, Siepmann T, Bohlmann F. *Justus Liebigs Ann Chem* 1963; **661**:1–37.

Dong DC, Winnik MA. *Can J Chem* 1984;**62**:2560–2565.

Dove MFA, Clifford AF. In: Jander G, Spandau H, Addison CC, editors. *Chemie in nichtwässrigen Lösungsmitteln*. Volume **I**, Braunschweig/Oxford: Vieweg/Pergamon; 1971. p 119–300 (in English).

Doye JPK, Wales DJ. *Phys Rev B* 1999;**59**:2292–2300.

Drago RS. *Pure Appl Chem* 1980;**52**:2261–2274.

Drago RS. *J Chem Soc Perkin Trans* 1992;**2**:2261–2274.

Drago RS, Kabler RA. *Inorg Chem* 1972;**11**:3144–3145.

Drago RS, Wayland BB. *J Am Chem Soc* 1965;**87**:3571–3577.

Drago RS, Wong N, Ferris DC. *J Am Chem Soc* 1991;**113**:1970–1977.

Drljaca A, Hubbard CD, van Eldrik R, Asano T, Basilevsky MV, le Noble WJ. *Chem Rev* 1998;**98**:2167–2289.

Drougard Y, Decroocq D. *Bull Soc Chim Fr* 1969:2972–2983.

Dubois J-E, Bienvenüe A. *J Chim Phys* 1968;**65**:1259–1265.

Dunn EJ, Buncel E. *Can J Chem* 1989;**67**:1440–1448.

Dunn WJ III, Wold S, Edlund U, Hellberg S, Gasteiger J. *Quant Struct Act Relat* 1984;**3**:1313–1317.

Dunn PJ, Galvin S, Hettenbach K. *Green Chem* 2004;**6**:43.

Durrell SR, Wallqvist A. *Biophys J* 1996;**71**:1696–1706.

Dutkiewicz M. *J Chem Soc Faraday Trans* 1990;**86**:2237–2241.

Dye JL. *Solutions métal-ammoniac: propriétés physicochimiques*. Lille/New York: Universite' Catholique/Benjamin; 1964. p 137–145.

Dyson PJ, Ellis DJ, Parker DG, Welton m T. *Chem Commun* 1999; (1):25–26.

Earle MJ, Katdare SP, Ramani A. In: Sanghi R, Srivastava MM, editors. *Green Chemistry: Environment Friendly Alternatives*. Pangbourne: Alpha Science; 2003. p 106–122.

Edwards JO. *J Am Chem Soc* 1954;**76**:1540–1547.

Edwards JO. *J Am Chem Soc* 1956;**78**:1819–1820.

Eigen M, Tamm K. *Z Electrochem* 1962;**66**:107–121.

El Hosary AA, Kerridge DH, Shams El Din AM. In: Inman D, Lovering DG, editors. *Ionic Liquids*. New York/London: Plenum; 1981. p 339–362.

Eliasson B, Johnels D, Wold S, Edlund U, Sjöström M. *Acta Chim Scand B* 1982; **36**:155–164.

Eliel EL, Hofer O. *J Am Chem Soc* 1973;**95**:8041–8045.

Eliel EE, Wilen SH, Mander LN. *Stereochemistry of Organic Compounds*. New York: Wiley-Interscience; 1994. p 416–421.

Ellis B, Keim W, Wasserscheid P. *Chem Commun* 1999; (4):337–338.

Endres F. *ChemPhysChem* 2002;**3**:144–154.

Erdey-Grúz T. *Transport Phenomena in Aqueous Solutions*. New York: Halstead; 1974.

Étard A. *Ann Chim Phys* 1881;**5, 22**:218–286.

Evans MG, Polanyi M. *Trans Faraday Soc* 1935;**31**:875–894.

Ewald P. *Ann Phys* 1921;**64**:253–287.

Eyring H. *J Chem Phys* 1935;**3**:107–115.

Fabre P-L, Devynck J, Tremillon B. *Chem Rev* 1982;**82**:591–614.

Fainberg AS, Winstein S. *J Am Chem Soc* 1956;**78**:2770–2777.

Fajans K. *Naturwiss* 1923;**11** (10):165–172.

Fajans K. *Chem Eng News* 1965;**43**:96.

Fang YR, Lai ZG, Westaway KC. *Can J Chem* 1998;**76**:758.

Feng D-F, Kevan L. *Chem Rev* 1980;**80**:1–20.

Flood H, Förland T, Motzfeldt K. *Acta Chim Scand* 1952;**6**:257–269.

Foresman JB, Frisch Æ. *Exploring Chemistry with Electronic Structure Methods: A Guide to Using Gaussian*. Pittsburgh (PA): Gaussian, Inc.; 1993.

Fowler HW, Fowler FG, Sykes JB. *Concise Oxford Dictionary*. 6th ed. Oxford: Oxford University Press; 1976.

Francisco M, van den Bruinhorst A, Kroon MC. *Angew Chem Int Ed* 2013;**52**:3074–3085.

Freemantle M. *Chem Eng News* 1998;**30**:32–37.

Freemantle M. *Chem Eng News* 2000;**79**:37–50.

Frémy ME. *Ann Chim Phys* 1856;**47**:5–50.

Fujii T, Hayashi R, Kawasaki S, Suzuki A, Oshima Y. *J Supercrit Fluids* 2011;**58** (1): 142–149.

Fuoss RM, Accascina F. *Electrolytic Conductance*. New York/London: Interscience; 1959.

Fuoss RM, Krauss CA. *J Am Chem Soc* 1933;**55**:1019–1028.

Gama S, Mackay D, Arnot JA. *Green Chem* 2012;**14**:1094–1102.

Gielen M, Nasielski J. *J Organomet Chem* 1967;**7**:273–280.

Gillespie RJ, Peel TE. *J Am Chem Soc* 1971;**93**:5083–5087.

Gillespie RJ, Peel TE. *J Am Chem Soc* 1973;**95**:5173–5178.

Gillespie RJ, Robinson EA. In: Waddington TC, editor. *Non-Aqueous Solvent Systems*. London: Academic Press; 1965. p 117.

Godfrey MB. *ChemTech* 1972;**2**:359–363.

Gold V. *J Chem Soc Perkin Trans* 1976;**2**:1531–1532.

Grant GH, Richards WG. *Computational Chemistry*. Oxford/New York/Toronto: Oxford University Press; 1995.

Gritzner G. *J Mol Liq* 1997;**73, 74**:487–500.

Gronwall TH, LaMer VK, Sandved K. *Z Phys* 1928;**29**:358–393.

Grunwald E, Winstein S. *J Am Chem Soc* 1948;**70**:841–846. 846–854.

Gu Y, Jérome F. *Chem Soc Rev* 2013;**42**:9550–9570.

Guillot B, Guissani Y. *J Chem Phys* 1993;**99**:8075–8094.

Guillot B, Guissani Y, Bratos S. *J Chem Phys* 1991;**95**:3643–3648.

Guterman L. *Chemistry*. Washington (DC): American Chemical Society; 1999. p 12–14.

Gutmann V. *Coord Chem Rev* 1967;**2**:239–256.

Gutmann V. *Coordination Chemistry in Non-Aqueous Solutions*. Vienna/New York: Springer-Verlag; 1968.

Gutmann V. *Chimia* 1969;**23**:285–292.

Gutmann V, Wychera E. *Inorg Nucl Chem Lett* 1966;**2**:257–260.

Haberfield P. *J Am Chem Soc* 1971;**93**:2091–2093.

Haberfield P, Lux MS, Rosen D. *J Am Chem Soc* 1977;**99**:6828–6831.

Hahn S, Miller WM, Lichtenthaler R, Prausnitz JM. *J Solution Chem* 1985;**14**:129–137.

Hall NF, Conant JB. *J Am Chem Soc* 1927;**49**:3047–3061. 3062–3070.

Hammett LP. *J Am Chem Soc* 1937;**59**:96–103.

Hammett LP. *Physical Organic Chemistry*. 2nd ed. New York: McGraw-Hill; 1970.

Hammett LP, Deyrup AJ. *J Am Chem Soc* 1932;**54**:2721–2739.

Hansch C. *Acc Chem Res* 1969;**2**:232–239.

Hansen JP, McDonald IR. *Theory of Simple Liquids*. London: Academic Press (Harcourt Brace Jovanovich); 1986.

Hantzsch A, Caldwell KS. *Z Phys Chem* 1908;**37**:227–240.

Hao C, March RE, Croley TR, Smith JC, Rafferty SP. *J Mass Spectrom* 2001;**36**:79–96.

Hao CH, March RE. *Int J Mass Spectrom* 2001;**212**:337–357.

Hartmann H, Schmidt AP. *Z Phys Chem* 1969;**66**:183.

Hashiguchi BG, Konnick MM, Bischof SM, Gustafson SJ, Devarajan D, Gunsalas N, Ess DH, Periana RA. *Science* 2014;**343**:1232–1237.

Hayes B. *Am Sci* 2013;**101** (2):92–97.

Headley AD, Ni B. Chiral imidazolium liquids: their synthesis and influence on the outcome of organic reactions. *Aldrichim Acta* 2007;**40** (4):107.

Hehre WJ, Radom L, Schleyer P v R, Pople JA. *Ab Initio Molecular Orbital Theory*. New York: John Wiley & Sons, Inc.; 1986.

Hildebrand JH, Carter JM. *Proc Natl Acad Sci* 1930;**16**:285–288.

Hildebrand JH, Scott RL. *The Solubility of Nonelectrolytes*. 3rd ed. New York: Reinhold; 1950.

Hildebrand JH, Scott RL. *Regular Solutions*. Englewood Cliffs (NJ): Prentice-Hall; 1962.

Hinton JF, Amis ES. *Chem Rev* 1971;**71**:627–674.

Hofer TS, Randolf BR, Rode BM. In: Leszczynski J, editor. *Solvation Effects on Molecules and Biomolecules*. New York: Springer Science+Business Media; 2008.

Hogen-Esch TE, Smid J. *J Am Chem Soc* 1966;**88**:307–318. 318–324.

Hohenberg P, Kohn W. *Phys Rev* 1964;**136B**:864–871.

Horvath IT, Rabai J. *Science* 1994;**266**:72–75.

House HO. *Modern Synthetic Reactions*. Menlo Park (CA): Benjamin; 1972.

Howarth J, Hanlon K, Fayne D, McCormac P. *Tetrahedron Lett* 1997;**38**:3097–3100.

Hughes ED, Ingold CK. *J Chem Soc* 1935:244–255.

Huheey JE, Keiter EA, Keiter RL. *Inorganic Chemistry: Principles of Structure and Reactivity*. 4th ed. New York: HarperCollins College Publishers; 1993.

Hunt JP. *Metal Ions in Solution*. New York: Benjamin; 1963. p 14–17 27–35.

Hurley FH, Weir JP. *J Electrochem Soc* 1951;**98**:203–206.

Ingold CK. *Structure and Mechanism in Organic Chemistry*. 2nd ed. Ithaca (NY): Cornell University Press; 1969.

Isaacs NS. In: Buncel E, Lee CC, editors. *Isotopes in Organic Chemistry*. Volume **6**, Amsterdam: Elsevier; 1984.

Jander J. In: Jander G, Spandau H, Addison CC, editors. *Chemie in nichtwässrigen ionisierenden Lösungsmitteln*. New York/London/Braunschweig: Vieweg/John Wiley & Sons, Inc.; 1966.

Jander J, Lafrenz C. *Ionizing Solvents*. Weinheim: Verlag Chemie; 1970.

Janowski A, Turowska-Tyrk I, Wrona PK. *J Chem Soc Perkin Trans* 1985;**2**:821–825.

Jencks WP. *Chem Rev* 1985;**85**:511–526.

Jessop PG. *Green Chem* 2011;**13**:1391–1398.

Jessop PG, Phan L, Carrier A, Robinson S, Dürr CJ, Harjani JR. *Green Chem* 2010; **12**:809–814.

Ji P, Atherton JH, Page MI. *Faraday Discuss* 2010;**145**:15–25.

Ji P, Atherton JH, Page MI. *Org Biomol Chem* 2012;**10**:5732–5739.

Johnson CD. *Chem Rev* 1975;**75**:755–765.

Johnson DE. *Applied Multivariate Methods for Data Analysts*. Pacific Grove (CA): Brooks/Cole; 1998.

Johnstone AR. *Chem Canada* 1978;**30** (November):33(letter).

Jolly WL. *The Synthesis and Characterization of Inorganic Compounds*. Englewood Cliffs (CA): Prentice-Hall; 1970.

Jorgensen WL. *Chem Phys Lett* 1982;**92**:405–410.

Kamlet MJ, Taft RW. *J Am Chem Soc* 1976;**98**:377–383. 2886–2894.

Kamlet MJ, Taft RW. *J Chem Soc Perkin Trans* 1979;**2**:337–341. 349–356.

Kamlet MJ, Abboud J-LM, Taft RW. *J Am Chem Soc* 1977;**99**:6027–6038.

Kamlet MJ, Jones ME, Taft RW. *J Chem Soc Perkin Trans* 1979;**2**:342–348.

Kamlet MJ, Abboud J-LM, Taft RW. *Prog Phys Org Chem* 1981;**13**:485–630.

Kamlet MJ, Abboud J-LM, Abraham MH, Taft RW. *J Org Chem* 1983;**48**:2877–2887.

Karabatsos GJ, Fenolio DJ. *J Am Chem Soc* 1969;**91**:1124–1129.

Kebarle P. In: Szwarc M, editor. *Ions and Ion Pairs in Organic Reactions*. Volume **1**, New York: John Wiley & Sons, Inc.; 1972.

Kerridge DH. In: Lagowski JJ, editor. *The Chemistry of Nonaqueous Solvents*. Volume **V**, New York: Academic Press; 1978. p 269–329.

Kettle SFA. *Physical Inorganic Chemistry*. Oxford: Spektrum; 1996. p 331–335.

Kier LB. *J Pharm Sci* 1981;**70**:930–933.

Kilpatrick M, Jones JG. In: Lagowski JJ, editor. *The Chemistry of Nonaqueous Solvents*. Volume **II**, New York: Academic Press; 1967. p 43–99.

Kilpatrick M, Lewis TJ. *J Am Chem Soc* 1956;**78**:5186–5189.

Kirkwood JG. *J Chem Phys* 1934;**2**:351–361.

Kleinberg R, Brewer P. *Am Sci* 2001;**89**:244–251.

Klopman G. *J Am Chem Soc* 1968;**90**:223–234.

Knauer BR, Napier JJ. *J Am Chem Soc* 1974;**98**:4395–4400.

Koga K, Tanaka H, Zeng XC. *J Phys Chem* 1996;**100**:16711–16719.

Kohn W, Sham LJ. *Phys Rev* 1965;**140 A**:1133–1138.

Koo IS, An SK, Yang K, Koh HJ, Choi MH, Lee I. *Bull Korean Chem Soc* 2001;**22** (8): 842–846.

Koppel IA, Paju A. *Org React* 1974;**11**:121.

Koppel IA, Pal'm VA. In: Chapman NB, Shorter J, editors. *Advances in Linear Free Energy Relationships*. New York/London: Plenum; 1972. p 203ff.

Kosower EM. *J Am Chem Soc* 1958;**80**:3253–3260.

Kosower EM. *An Introduction to Physical Organic Chemistry*. New York: John Wiley & Sons, Inc.; 1968.

Kováts E. *Z Anal Chem* 1961;**181**:351–369.

Kováts E. *Adv Chromatogr* 1965;**1**:229–247.

Kraus CA, Brey WC. *J Am Chem Soc* 1913;**35**:1315–1434.

Kresge AJ, More O'Ferrall RA, Powell FM. In: Buncel E, Lee CC, editors. *Isotopes in Organic Chemistry*. Volume **7**, Amsterdam: Elsevier; 1987. p 177–273.

Kritzer P, Dinjus E. *Chem Eng J* 2001;**83** (3):207–214.

Kumar A. *Chem Rev* 2001;**101**:1–19.

Kusalik PG, Svishchev I. *Science* 1994;**265**:1219–1221.

Kwon DS, Lee GH, Um IH. *Bull Korean Chem Soc* 1989;**10**:620–621.

Kyba EP, Koga K, Sousa LR, Siegel MG, Cram DJ. *J Am Chem Soc* 1973;**95**:2692–2693.

Lagowski JJ. *Pure Appl Chem* 1971;**25**:429–464.

Lagowski JJ, Moczygemba GA. In: Lagowski JJ, editor. *The Chemistry of Nonaqueous Solvents*. Volume **II**, New York: Academic Press; 1967. p 319–371.

Laidler KJ. *Chemical Kinetics*. 3rd ed. New York: Harper and Row; 1987.

Laidler KJ, Meiser JM. *Physical Chemistry*. 2nd ed. Boston (MA): Houghton Mifflin; 1995.

Langford CH, Tong JPK. *Acc Chem Res* 1977;**10**:258–264.

Langhals H. *Nouv J Chim* 1982a;**6**:265–267.

Langhals H. *Angew Chem Int Ed Engl* 1982b;**B21**:724–733.

Latimer WM, Pitzer KS, Slansky CM. *J Chem Phys* 1939;**7**:108–111.

Lau YK, Ikuta S, Kebarle P. *J Am Chem Soc* 1982;**104**:1462–1469.

Laurence C, Legros J, Nicolet P, Vuluga D, Chantzis A, Jacquemin D. *J Phys Chem B* 2014;**118**:7594–7608.

Leach AR. *Molecular Modelling: Principles and Applications*. Harlow: Addison Wesley Longman; 1996.

Lennard-Jones JE. *Proc Roy Soc Lond A* 1924;**106** (738):463–477.

Leo A, Hansch C, Elkins D. *Chem Rev* 1971;**71**:525–616.

Leo A. *J Chem Soc Perkin Trans* 1983;**2**:825–838.

Levine IN. *Quantum Chemistry*. 7th ed. Englewood Cliffs (NJ): Prentice Hall; 2013.

Liu CT, Maxwell CI, Edwards DR, Neverov AA, Mosey NJ, Brown RS. *J Am Chem Soc* 2010;**132**:16599–16609.

Lux H. *Z Elektrochem* 1939;**45**:303–309.

Mackay D, Shiu WY, Ma KC. *Illustrated Handbook of Physical-Chemical Properties and Environmental Fate of Organic Chemicals*. Boca Raton (FL): Lewis; 1992.

Madan B, Sharp K. *J Phys Chem* 1996;**100**:7714–7721.

Maksimović ZB, Reichardt C, Spirić AZ. *Z Anal Chem* 1974;**270**:100–104.

Malinowski ER, Howery DG. *Factor Analysis in Chemistry*. New York: John Wiley & Sons, Inc. reprinted 1989 by Krieger, Malabar (FL); 1980.

Manuta DM, Lees AJ. *Inorg Chem* 1986;**25**:3212–3218.

Marcus Y. *Pure Appl Chem* 1983;**55**:977–1021.

Marcus Y. *Ion Solvation*. Chichester: John Wiley & Sons, Inc.; 1985.

Marcus Y. *J Phys Chem* 1987;**91**:4422–4428.

Marcus Y. *Chem Soc Rev* 1993;**22**:409–416.

Marcus Y. *The Properties of Solvents*. Chichester: John Wiley & Sons, Inc.; 1998.

Maria P-C, Gal J-F. *J Phys Chem* 1985;**89**:1296–1304.

Maria P-C, Gal J-F, de Franceschi J, Fargin E. *J Am Chem Soc* 1987;**109**:483–492.

Marziano NC, Cimino GM, Passerini RC. *J Chem Soc Perkin Trans* 1973;**2**:1915–1922.

Marziano NC, Traverso PG, Tomasin A, Passerini RC. *J Chem Soc Perkin Trans* 1977;**2**: 309–313.

Mashima M, McIver RR, Taft RW, Bordwell FG, Olmstead WN. *J Am Chem Soc* 1984;**106**:2717–2718.

Matubayasi N, Levy RM. *J Phys Chem* 1996;**100**:2681–2688.

Matyushov DV, Schmid R, Ladanyi BM. *J Phys Chem* 1997;**101**:1035–1050.

Maurois A. *The Silence of Colonel Bramble*. London: The Bodley Head; 1927. p 176.

Mayer U, Gutmann V, Gerger W. *Monatsh Chem* 1975;**106**: 1235–1257.

Mayer U, Gerger W, Gutmann V. *Monatsh Chem* 1977;**108**: 489–498.

Melander L, Saunders WH. *Reaction Rates of Isotopic Molecules*. New York: John Wiley & Sons, Inc.; 1980.

Meng EC, Kollman PA. *J Phys Chem* 1996;**100**:11460–11470.

Menschutkin N. *Z Phys Chem* 1887;**1**:611.

Menschutkin N. *Z Phys Chem* 1890;**6**:41.

Metropolis N, Rosenbluth AW, Rosenbluth MN, Teller AH, Teller E. *J Chem Phys* 1953; **21**:1087–1092.

Meyer KH, Hopff H. *Ber Dtsch Chem Ges* 1921;**54**:579–580.

Miertus S, Scrocco E, Tomasi J. *Chem Phys* 1981;**55**:117–129.

Milne AA. *Winnie the Pooh*. New York: Dutton; 1926.

Mitchell SA. In: Fontijn A, editor. *Gas-Phase Metal Reactions*. Amsterdam: Elsevier; 1992. p 227–252.

Møller C, Plesset MS. *Phys Rev* 1934;**46**:618–622.

Moore WJ. *Physical Chemistry*. 4th ed. Englewood Cliffs (NJ): Prentice-Hall; 1972. p 571.

More O'Ferrall RA, Koeppl GW, Kresge AJ. *J Am Chem Soc* 1971;**93**:9–20.

Morrison RT, Boyd RN. *Organic Chemistry*. 6th ed. Englewood Cliffs (NJ): Printice-Hall; 1992.

Mu L, Drago RS, Richardson DE. *J Chem Soc Perkin Trans* 1998;**2**:159–167.

Muddana HS, Sapra NV, Fenley AT, Gilson MK. *J Chem Phys* 2013;**138**:224504.

Mukerjee P, Ramachandran C, Pyter RA. *J Phys Chem* 1982;**86**:3189–3197.

Nash O. *I'm a Stranger Here Myself*. Boston (MA): Little, Brown & Co.; 1938.

Némethy G, Scheraga HA. *J Chem Phys* 1962;**36**:3401–3417.

Newton MD. *J Phys Chem* 1997;**67**:5535–5546.

Nigretto J-M, Jozefowicz M. In: Lagowski JJ, editor. *The Chemistry of Nonaqueous Solvents*. Volume **V**, New York: Academic Press; 1978. p 179–250.

Ogston AG. *Trans Faraday Soc* 1936;**32**:1679–1691.

Oh YH, Jang GG, Lim GT, Ryu ZH. *Bull Korean Chem Soc* 2002;**23**:1089–2429.

Ohno H, editor. *Electrochemical Aspects of Ionic Liquids*. 2nd ed. New York: John Wiley & Sons, Inc.; 2011.

Okazaki S, Nakanishi K, Touhara H, Adachi Y. *J Chem Phys* 1979;**71**:2421–2429.

Okazaki S, Nakanishi K, Touhara H, Watanabi N, Adachi Y. *J Chem Phys* 1981; **74**:5863–5871.

Olah GA, Prakash GKS, Sommer J. *Superacids*. New York: John Wiley & Sons, Inc.; 1985.

Olivier-Bourbigou H, Magna L, Morvan D. *Appl Catal Gen* 2010;**373**:1–56.

Olmstead WN, Brauman JI. *J Am Chem Soc* 1977;**99**:4219–4228.

Onsager L. *J Am Chem Soc* 1936;**58**:1486–1493.

Oshima T, Nagai T. *Tetrahedron Lett* 1985;**26**:4785–4788.

Owen RB, Waters GW. *J Am Chem Soc* 1938;**60**:2371–2379.

Parker AJ. *Q Rev* 1962;**16**:163–187.

Parker AJ. *Adv Phys Org Chem* 1967;**5**:173–235.

Parker AJ. *Chem Rev* 1969;**69**:1–32.

Peacock SC, Cram DJ. *J Chem Soc Chem Commun* 1976:282–284.

Pearson RG. *J Am Chem Soc* 1963;**85**:3533–3539.

Pearson RG, editor. *Hard and Soft Acids and Bases*. Stroudsberg (PA): Dowden, Hutchinson & Ross; 1973.

Pearson RG. *Inorg Chem* 1988;**27**:734–741.

Pellerite MJ, Brauman JI. In: Buncel E, Durst T, editors. *Comprehensive Carbanion Chemistry*. Amsterdam: Elsevier; 1980.

Percus JK, Yevick GJ. *Phys Rev* 1958;**110**:1–13.

Persson I, Sandström M, Goggin PL. *Inorg Chim Acta* 1987;**129**:183–197.

Pethybridge AD, Prue JE. *Prog Inorg Chem* 1972;**17**:327–390.

Pine SH. *Organic Chemistry*. 5th ed. New York: McGraw-Hill; 1987.

Plešek J, Heřmaňek S. In: Mayer K, trans. *Sodium Hydride: Its Use in the Laboratory and in Technology*. Cleveland (OH): CRC Press; 1968.

Pollett P, Eckert CA, Liotta CL. *Chem Sci* 2011;**2**:609–614.

Poos GI, Arth GE, Beyler RE, Sarett LH. *J Am Chem Soc* 1953;**75**:422–429.

Porterfield WW, Yoke JT. Inorganic compounds with unusual properties. *Adv Chem* 1976;**150**:104–111.

Pourbaix MJN. *Thermodynamics of Dilute Aqueous Solutions*. London: Arnold; 1949.

Pourbaix MJN, Van Muylder J, de Zhoukov N. *Atlas d'Equilibres électroniques à 25°C*. Paris: Gauthier-Villars; 1963.

Pregel MJ, Dunn EJ, Negelkerke R, Thatcher GRJ, Buncel E. *Chem Soc Rev* 1995;**24**:449–455.

Rabinowitch E, Wood WC. *Trans Faraday Soc* 1936;**32**:1381–1387.

Ràfols C, Rosés M, Bosch E. *J Chem Soc Perkin Trans* 1997;**2**:234–248.

Rahimi AK, Popov AI. *Inorg Nucl Chem Lett* 1976;**12**:703–707.

Rau H. *Chem Rev* 1983;**83**:535–547.

Rauhut G, Clark T, Steinke T. *J Am Chem Soc* 1993;**115**:9174–9181.

Reichardt C. *Justus Liebigs Ann Chem* 1971;**752**:64–67.

Reichardt C. *Solvents and Solvent Effects in Organic Chemistry*. 2nd ed. Weinheim: VCH; 1988.

Reichardt C. *Chem Rev* 1994;**94**:2319–2358.

Reichardt C. *Solvents and Solvent Effects in Organic Chemistry*. 3rd ed. Weinheim: Wiley-VCH; 2003.

Reichardt C, Harbusch-Görnert E. *Justus Liebigs Ann Chem* 1983:721–743.

Reichardt C, Welton T. *Solvents and Solvent Effects in Organic Chemistry*. 4th ed. Weinheim: Wiley-VCH; 2011.

Reichardt C, Löbbecke S, Mehranpour AM, Schäfer G. *Can J Chem* 1998;**76**:686–694.

Ritchie CD. In: Coetzee JF, Ritchie CD, editors. *Solute-Solvent Interactions*. New York/London: Marcel Dekker; 1969.

Robinson RA, Stokes RH. *Electrolyte Solutions*. London: Butterworths; 1959.

Röllgen FW, Bramer-Wegner E, Buttering L. *J Phys Colloq* 1984;**45** (Suppl. 12):C9.

Romeo R, Minniti D, Lanza S. *Inorg Chem* 1980;**19**(12): 3663–3668.

Rotzinger RP. *Chem Rev* 2005;**105**:2003–2038.

Rusz C, König B. *Green Chem* 2012;**14**(11): 2969–2982.

Sahu S. In: Sanghi R, Srivastava MM, editors. *Green Chemistry: Environment Friendly Alternatives*. Pangbourne: Alpha Science; 2003. p 122–145.

Scatchard G. *Chem Rev* 1931;**8**:321–333.

Scatchard G. *Chem Rev* 1932;**10**:229–240.

Scott JL. In: Sanghi R, Srivastava MM, editors. *Green Chemistry: Environment Friendly Alternatives*. Pangbourne: Alpha Science; 2003. p 55–72.

Scriven EFV, Toomey JE, Murugan R. *Kirk-Othmer, Encyclopedia of Chemical Technology*. 4th ed. Volume **20**, 1994. p 641–679.

Shafirovitch V, Dourandin A, Geacintov NE. *J Phys Chem B* 2001;**105**:8431–8435.

Shaik SS, Schlegel HB, Wolfe S. *Theoretical Aspects of Physical Organic Chemistry: The $S_N2$ Mechanism*. New York: John Wiley & Sons, Inc.; 1992.

Simkin BY, Sheikhet II. In: Kemp TJ, editor. *Quantum Chemical and Statistical Theory of Solutions: A Computational Approach*. Engl. ed. London: Ellis Horwood; 1995.

Sjöström M, Wold S. *J Chem Soc Perkin Trans* 1981;**2**:104–109.

Skoog DA, West DM, Holler FJ. *Fundamentals of Analytical Chemistry*. New York: Saunders; 1989. p 124–130.

Smith H. In: Jander G, Spandau H, Addison CC, editors. *Chemie in nichtwässrigen ionisieren-den Lösungsmitteln*. Braunschweig/New York/London: Vieweg/John Wiley & Sons, Inc./Interscience; 1963.

Smith MB, March J. *March's Advanced Organic Chemistry*. Hoboken, NJ: Wiley-Interscience; 2007.

Smith RA. *Kirk-Othmer, Encyclopedia of Chemical Technology*. 4th ed. Volume **11**, 1994. p 355–376.

Smithson JM, Williams RJP. *J Chem Soc* 1958:457.

Solomons G, Fryhle C. *Organic Chemistry*. 7th ed. New York: John Wiley & Sons, Inc.; 2000.

Song CE, Shim WH, Roh EJ, Choi JH. *Chem Commun* 2000; (17):1695–1696.

Song CE, Shim WH, Roh EJ, Lee S, Choi JH. *Chem Commun* 2001; (12):1122–1123.

Stairs RA. *J Chem Phys* 1957;**27**:1431–1432.

Stairs RA. *Can J Chem* 1962;**40**:1656–1659.

Stairs RA. In: Furter WF, editor. *Thermodynamic Behaviour of Electrolytes in Mixed Solvents*. Volume **155**, Washington (DC): American. Chemical Society; 1976. Advances in Chemistry Series; p 332–342.

Stairs RA. *Chem Eng News* 1978a;June:36(letter).

Stairs RA. *Chem Canada* 1978b;**30** (September):31(letter).

Stairs RA. In: Furter WF, editor. *Thermodynamic Behaviour of Electrolytes in Mixed Solvents-II*. Volume **177**, Washington (DC): American Chemical Society; 1979. Advances in Chemistry Series; p 167–176.

Stairs RA. *Chem 13 News*. Canada: University of Waterloo; 1983. p 11–12.

Stairs RA, Buncel E. *Can J Chem* 2006;**84**:1580–1591.

Stein G, Würzberg E. *J Chem Phys* 1975;**62**:208–213.

Stewart R. *The Proton: Applications to Organic Chemistry*. Orlando (FL): Academic Press; 1985.

Stillinger FH, Rahman A. *J Chem Phys* 1974;**60**:1545–1557.

Strehlow H, Schneider H. *J Chim Phys* 1969;**66**:118–123.

Strehlow H, Schneider H. *Pure Appl Chem* 1971;**25**:327–344.

Streitwieser A, Heathcock CH, Kosower EM. *Introduction to Organic Chemistry*. New York: Macmillan; 1992.

Suarez PAZ, Dullius JEL, Einloft S, de Souza RF, Dupont J. *Inorg Chim Acta* 1997;**255**:207–209.

Sugiyama H, Fischer U, Hungerbühler J. 2006. The EHS Tool. ETH Zurich, Safety & Environmental Technology Group, Zurich. Available at http//www.sust-chem.ethz.ch/tools/EHS. Accessed February 11, 2015.

Sumimoto M, Iwane N, Takahama T, Sakaki S. *J Am Chem Soc* 2004;**126**:10457–10471.

Svishchev I, Kusalik PG, Wang S, Boyd RJ. *J Chem Phys* 1996;**105**:4742–4750.

Swain CG, Swain MS, Powell AL, Alunni S. *J Am Chem Soc* 1983;**105**:502–513.

Swieton G, Jouanne Jv, Kelm H, Huisgen R. *J Chem Soc Perkin Trans II* 1983:37–43.

Symons EA, Buncel E. *J Am Chem Soc* 1972;**94**:3641–3642.

Symons EA, Clermont MJ. *J Am Chem Soc* 1981;**103**:3127–3130.

Symons EA, Clermont MJ, Coderre LA. *J Am Chem Soc* 1981;**103**:3131–3135.

Szwarc M. *Carbanions, Living Polymers and Electron Transfer Processes*. New York: Wiley-Interscience; 1968. p 216–225.

Szwarc M, editor. *Ions and Ion Pairs in Organic Reactions*. New York: Wiley-Interscience; 1972.

Taber GP, Pfisterer DM, Colberg JC. *Org Process Res Dev* 2004;**8**:385.

Taft RW, Bordwell FG. *Acc Chem Res* 1988;**21**:463–469.

Taft RW, Abboud J-LM, Kamlet MJ. *J Am Chem Soc* 1981;**103**:1080–1086.

Tamura H, Yamasaki H, Sato H, Sakaki S. *J Am Chem Soc* 2003;**125**:16114–16126.

Tanaka H. *J Chem Phys* 1987;**86**:1512–1520.

Tang S, Baker GA, Zhao H. *Chem Soc Rev* 2012;**41**:4030–4066.

Tapia O, Goscinski O. *Mol Phys* 1975;**29**:1653–1661.

Theilemans A, Massart DI. *Chimia* 1985;**39**:236–243.

Thomson BA, Iribarne JV. *J Chem Phys* 1979;**71**:4451–4463.

Tobe ML, Burgess J. *Inorganic Reaction Mechanisms*. Harlow: Addison Wesley Longman; 1999.

Trémillon B. *Pure Appl Chem* 1971;**25**:395–428.

Trémillon B. In: Corcoran N, trans. *Chemistry in Non-Aqueous Solvents*. Dordrecht/Boston (MA): Reidel; 1974.

Troe J. *Annu Rev Phys Chem* 1978;**29**:223–250.

Um I-H, Buncel E. *J Am Chem Soc* 2001;**123**:11111–11112.

Um I-H, Yoon HW, Kwon DS. *Bull Korean Chem Soc* 1993;**14**:425–427.

Um I-H, Shin Y-H, Han JY, Buncel E. *Can J Chem* 2006;**84** (11):1550–1556.

Usanovich M. *Zh Obshch Khim* 1939;**9**:182–192.

Valdez-Rojas JE, Rios-Guerra H, Ramirez-Sánchez AL, Garcia-González G, Alvarez-Toledano C, López-Cortés JG, Toscano RA, Penieres-Carrillo JG. *Can J Chem* 2012;**90**(7):567–573.

van Eldik R, Hubbard CD, editors. *Chemistry under Extreme or Non-Classical Conditions*. New York/Heidelberg: John Wiley/SpektrumChapters 2–4; 1997.

van Eldik R, Meyerstein D. *Acc Chem Res* 2000;**33**:207–214.

van Eldik R, Asano T, le Noble WJ. *Chem Rev* 1989;**89**:549–688.

van Leeuwen JMJ, Groeneveld J, deBoer J. *Physica* 1959;**25**:792–808.

Vinci D, Donaldson M, Hallett JP, John EA, Pollett P, Thomas CA, Grilly JD, Jessop PG, Liotta CL, Eckert CA. *Chem Commun* 2007; (14):1427–1429.

Walden p. *Bull Acad Imp Sci* 1914;**8**:405–422.

Walter W, Bauer OH. *Justus Liebigs Ann Chem* 1977:421–429.

Walther D. *J Prakt Chem* 1974;**316**:604–614.

Wang P, Nishimura D, Komatsu T, Kobira K. *J Supercrit Fluids* 2011;**58** (3):260–364.

Wasserscheid P, Welton T. *Ionic Liquids in Synthesis*. 2nd ed. New York: John Wiley & Sons, Inc.; 2008.

Wasserscheid P, Gordon CM, Hilgers C, Muldoon MJ, Dunkin IR. *Chem Commun* 2001; (13):1186–1187.

Welton T. *Chem Rev* 1999;**99**:2071–2083.

Werblan L, Rotowska A, Minc S. *Electrochim Acta* 1971;**16**:41–49.

Wheeler C, West KN, Liotta CL, Eckert CA. *Chem Commun* 2001; (10):887–888.

Wiberg KB. *Physical Organic Chemistry*. New York: John Wiley & Sons, Inc.; 1964. p 401.

Wilson KR. In: Moreau M, Turq P, editors. *Chemical Reactivity in Liquids: Fundamental Aspects*. New York: Plenum; 1989.

Winstein S, Grunwald E, Jones HW. *J Am Chem Soc* 1951;**73**:2700.

Winstein S, Clippinger E, Fainberg AH, Robinson GC. *J Am Chem Soc* 1954;**76**:2597–2598.

Winstein S, Fainberg AH, Grunwald E. *J Am Chem Soc* 1957;**79**:4146–4155.

Wold S, Sjöström M. In: Shorter J, Chapman NB, editors. *Correlation Analysis in Chemistry: Recent Advances*. New York: Plenum; 1978.

Wong MW, Wiberg KB, Frisch MJ. *J Am Chem Soc* 1992;**114**:1645–1652.

Wood JM, Hinchcliffe PS, Laws AP, Page MI. *J Chem Soc Perkin Trans* 2002;**2**:938–946.

Wooley EM, Hepler LG. *Anal Chem* 1972;**44**:1520–1523.

Wypych G, editor. *Handbook of Solvents*. Toronto/Norwich (NY): ChemTec/Wm. Andrew; 2001.

Yingst A, McDaniel DH. *Inorg Chem* 1967;**6**:1067–1068.

Yu H-A, Pettitt BM, Karplus M. *J Am Chem Soc* 1991;**113**:2425–2434.

Zhang SJ, Sun N, He XZ, Lu XM, Zhang XP. *J Phys Chem Ref Data* 2006;**35** (4):1476–1517.

Zhang Y, Bakshi BR, Demessie ES. *Environ Sci Technol* 2008;**42**:1724–1730.

# INDEX

Note: Page numbers in **bold** refer to Tables.

*Solvent Effects in Chemistry*, Second Edition. Erwin Buncel and Robert A. Stairs.
© 2016 John Wiley & Sons, Inc. Published 2016 by John Wiley & Sons, Inc.